# Occurrence and Analysis of Organometallic Compounds in the Environment

# Occurrence and Analysis of Organometallic Compounds in the Environment

**T. R. CROMPTON**
*Consultant, Anglesey, Gwynedd, Wales*

JOHN WILEY & SONS
Chichester · New York · Weinheim · Brisbane · Singapore · Toronto

Copyright © 1998 by John Wiley & Sons Ltd,
Baffins Lane, Chichester,
West Sussex PO19 1UD, England

National 01243 779777
International (+44) 1243 779777

e-mail (for orders and customer service enquiries): cs-books@wiley.co.uk.
Visit our Home Page on http://www.wiley.co.uk
or http://www.wiley.com

All rights reserved. No part of this publication may be reproduced, stored in a retrieval system, or transmitted, in any form or by any means, electronic, mechanical, photocopying, recording, scanning or otherwise, except under the terms of the Copyright, Designs and Patents Act 1988 or under the terms of a licence issued by the Copyright Licensing Agency, 90 Tottenham Court Road, London, UK W1P 9HE, without the permission in writing of the publisher.

*Other Wiley Editorial Offices*

John Wiley & Sons, Inc., 605 Third Avenue,
New York, NY 10158-0012, USA

WILEY-VCH Verlag GmbH, Pappelallee 3,
D-69469 Weinheim, Germany

Jacaranda Wiley Ltd, 33 Park Road, Milton,
Queensland 4064, Australia

John Wiley & Sons (Asia) Pte Ltd, 2 Clementi Loop #02-01,
Jin Xing Distripark, Singapore 129809

John Wiley & Sons (Canada) Ltd, 22 Worcester Road,
Rexdale, Ontario M9W 1L1, Canada

*Library of Congress Cataloging-in-Publication Data*

Crompton, T.R. (Thomas Roy)
   Occurrence and analysis of organometallic compounds in the environment / T.R. Crompton.
     p. cm.
   Includes bibliographical references (p.    ) and index.
   ISBN 0 471 97607 5 (hardbound : alk. paper)
   1. Organometallic compounds—Environmental aspects. I. Title.
QH545.074C76  1998
577.27'8—dc21                                       97-28546
                                                                       CIP

*British Library Cataloguing in Publication Data*

A catalogue record for this book is available from the British Library

ISBN 0 471 97607 5

Typeset in 10/12pt Times by Laser Words, Madras, India

# Contents

| | |
|---|---|
| **Preface** | **ix** |
| **1 Introduction** | **1** |
| **2 Concentrations of Organometallic Compounds in the Ecosystem** | **5** |
|    2.1 The River Ecosystem | 5 |
|    2.2 Coastal Water Ecosystem | 6 |
|    2.3 Open Seawater Ecosystem | 10 |
|    2.4 Fish and Other Creatures in the Ecosystem | 20 |
|          2.4.1 Fish | 22 |
|          2.4.2 Crustacea | 25 |
|    2.5 The Phytoplankton and Weed Ecosystem | 26 |
|    2.6 The Soil and Crops Ecosystem | 28 |
|          2.6.1 Consequences of Applying Sewage Sludge to Land | 29 |
|    2.7 Potable Water | 30 |
|    2.8 The Atmospheric Ecosystem | 31 |
|    References | 32 |
| **3 Toxic Effects of Organometallic Compounds** | **35** |
|    3.1 Toxicity Data | 35 |
|          3.1.1 Introduction | 35 |
|          3.1.2 Organomercury Compounds | 35 |
|          3.1.3 Organolead Compounds | 40 |
|          3.1.4 Organotin Compounds | 42 |
|          3.1.5 Organoarsenic Compounds | 48 |
|    3.2 Toxic Levels in the Human Duct | 49 |
|          3.2.1 Inorganic Elements | 49 |
|          3.2.2 Organometallic Compounds | 54 |
|    References | 56 |
| **4 Toxicity Testing** | **61** |
|    4.1 Test Details | 61 |
|    4.2 Effect of Test Substance Concentration on $LC_{50}$ and $LE_{50}$ Determinations on Fish and Crustacea | 62 |

4.3 Effect of Exposure Time on $LC_{50}$ and $LE_{50}$ Determinations on Fish and Crustacea .......................... 62
4.4 Effect of Other Experimental Parameters on $LC_{50}$ and $LE_{50}$ Determinations ............................... 64
4.5 Derivation of Maximum Safe Concentration Standard ($S$) of a Test Substance for Continuous Exposure to Fish ...... 64
4.6 Cumulative $LC_{50}$ Values ....................... 68
References ...................................... 73

## 5 Bioaccumulation Processes ........................... 75
5.1 Introduction ................................. 75
5.2 Bioaccumulation of Inorganic Metal Ions ........... 79
5.3 Bioaccumulation of Organometallic Compounds ....... 83
    5.3.1 Organolead Compounds .................. 83
    5.3.2 Organomercury Compounds ............... 83
    5.3.3 Organotin Compounds .................... 84
5.4 General Conclusions ........................... 85
    5.4.1 Fish .................................. 88
    5.4.2 Invertebrates .......................... 88
    5.4.3 Plants ................................ 88
References ...................................... 89

## 6 Analysis of Organometallic Compounds in the Environment . 91
6.1 Organomercury Compounds ....................... 91
    6.1.1 Natural and Potable Waters ............... 91
    6.1.2 Sea Waters ........................... 105
    6.1.3 Rainwater ............................ 110
    6.1.4 Industrial Effluents and Wastewaters ........ 110
    6.1.5 Sewage Effluents ...................... 112
    6.1.6 Soils and River Sediments ............... 113
    6.1.7 Plants, Crops and Aquatic Organisms ......... 119
    6.1.8 Biological Materials .................... 120
    6.1.9 In Air ............................... 133
    6.1.10 Preservation of Mercury Containing Samples .... 135
6.2 Organolead Compounds ........................ 139
    6.2.1 Natural and Potable Waters ............... 139
    6.2.2 Seawater ............................. 150
    6.2.3 Rainwater and Snow .................... 150
    6.2.4 Industrial Effluents and Wastewaters ......... 151
    6.2.5 Sediments ............................ 152
    6.2.6 Biological Materials .................... 153

|  |  |  | |
|---|---|---|---|
| | 6.2.7 | Plant Materials | 158 |
| | 6.2.8 | Air | 158 |
| 6.3 | Organotin Compounds | | 165 |
| | 6.3.1 | Natural and Potable Waters | 165 |
| | 6.3.2 | Seawater | 178 |
| | 6.3.3 | Rainwater | 182 |
| | 6.3.4 | Sediments | 182 |
| | 6.3.5 | Sewage Sludge | 184 |
| | 6.3.6 | Biological Materials | 184 |
| | 6.3.7 | Crops and Plants | 186 |
| 6.4 | Organocadmium Compounds | | 187 |
| 6.5 | Organoarsenic Compounds | | 187 |
| | 6.5.1 | Natural and Potable Waters | 187 |
| | 6.5.2 | Sea Water | 198 |
| | 6.5.3 | Sediments and Soils | 198 |
| | 6.5.4 | Industrial Effluents and Waste Waters | 199 |
| | 6.5.5 | Biological Materials | 202 |
| | 6.5.6 | Plants and Crops | 203 |
| 6.6 | Organoantimony Compounds | | 204 |
| | 6.6.1 | Natural Waters | 204 |
| 6.7 | Organoselenium Compounds | | 204 |
| | 6.7.1 | Natural Waters | 204 |
| | 6.7.2 | Biological Materials | 205 |
| 6.8 | Organogermanium Compounds | | 205 |
| | 6.8.1 | Natural and Organic Waters | 205 |
| 6.9 | Organomanganese Compounds | | 206 |
| | 6.9.1 | Natural Waters | 206 |
| | 6.9.2 | Air | 207 |
| 6.10 | Organocopper Compounds | | 207 |
| | 6.10.1 | Natural Waters | 207 |
| 6.11 | Organonickel Compounds | | 208 |
| | 6.11.1 | Air | 208 |
| 6.12 | Organosilicon Compounds | | 209 |
| | 6.12.1 | Natural Waters | 209 |
| | 6.12.2 | Biological Materials | 210 |
| 6.13 | Organoboron Compounds | | 210 |
| | 6.13.1 | Tissues | 210 |
| 6.14 | Organophosphorous Compounds | | 211 |
| | 6.14.1 | Environmental Samples | 211 |
| 6.15 | Organosulfur Compounds | | 211 |

| | | |
|---|---|---|
| 6.16 | Selection of Appropriate Analytical Methods | **211** |
| | 6.16.1 Water Samples | **211** |
| | 6.16.2 Sediments | **219** |
| | 6.16.3 Fish and Invertebrates | **220** |
| **References** | | **221** |
| **Index** | | **243** |

# Preface

A knowledge of the chemical structure and concentration of organometal compounds throughout the ecosystem is important in working out the pathways and mechanisms by which metals distribute themselves throughout the environment. This is especially so when it is realised that many organometallic compounds are considerably more toxic to sea creatures, animals and man than are the parent metals from which they are derived. This, in turn, has implications for the continued survival of sea creatures such as fish, crustacea, seals, land based creatures such as cattle, polar bears, reindeer and birds, and man who is higher up in the food chain. The presence of alkyl compounds of lead, mercury, tin and arsenic in the tissues of fish and crustacea is a proven fact and these are all ingested by man in his diet.

In many cases, regulations exist for maximum permitted concentrations of these toxic substances in food and in the amounts of such foods that it is advisable to eat in a stipulated period of time.

It is the aim of this book to review all aspects of the occurence, toxicity and analysis of organometallic compounds that occur in environmental samples. The principal elements discussed are mercury, lead, tin and arsenic and most recently cadmium. Other organometallic compounds which are discussed and occur in the environment less frequently and at lower concentrations include antimony, selenium, germanium, manganese, copper and nickel.

In Chapter 2 are discussed typical examples of the concentrations that occur of the above types of compounds in environmental samples such as natural water (rivers, ponds etc), snow, rainwater, coastal waters, open seawaters, river and ocean sediments, fish, crustacea, phytoplankton and plants that occur in rivers and oceans. The concentrations of organometallic compounds that occur in other types of environmental samples as a result of the permeation of these compounds into the ecosystem is also discussed. These include plants, crops, soil, potable water and the atmosphere.

The toxic effects of organometallic compounds on fish, crustacea, animals and man is discussed in Chapter 3 whilst in Chapter 4 is discussed the methodology for determining the toxic effects of organometallic compounds on the above species, including discussion on the measurement of $LC_{50}$, $LE_{50}$ and maximum safe concentration standards for continuous exposure.

In Chapter 5 the phenomenon of bioaccumulation of organometallic compounds and the corresponding inorganic metal ions in environmental sediments, plants, fish and crustacea and the consequences of this are discussed in detail.

Chapter 6 is devoted to a discussion of the analytical methodology employed to determine the types of organometallic compounds present in environmental samples. A wide variety of analytical techniques have been adapted to this end as shown in the table below. Obviously, the main thrust of analytical effort has been directed at the determination of the organometallic compounds of mercury, lead, tin and arsenic in all types of environmental samples. The determination of organomanganese and organonickel compounds originating from automobile exhausts in the atmosphere is disussed, as is the development of methods for determining organocompounds of antimony, selenium, germanium, manganese, copper, nickel and silicon in natural waters.

Gas chromatography, high performance liquid chromatography and atomic absorption spectrometry have been extensively used to determine organometallic compounds. As well as the conventional detectors used in the gas chromatographic and high performance liquid chromatographic methods, other more recently developed detector systems including inductively coupled plasma, atomic emission spectrometric and mass spectrometric methods have been used. Reduction of organometallic compounds to metal hydrides followed by atomic absorption spectrometry or gas chromatography has also been employed in the case of organometallic compounds of arsenic, tin and germanium. Techniques such as neutron activation analysis, nuclear magnetic resonance spectroscopy, x-ray fluorescence spectroscopy and chemical ionisation mass spectrometry have found limited applications to date.

Sample preservation techniques are discussed and also analytical preconcentration procedures. The latter procedure, in which the organometallic compound present in a large quantity of sample is concentrated into a smaller quantity before the analysis proper is commenced, enables detection limits to be reduced, a requirement that is often essential in environmental analysis.

|    | Natural water | Effluents | Sediments | Plant materials | Biological materials | Air |
|----|---------------|-----------|-----------|-----------------|----------------------|-----|
| Hg | a,c,i,k,l,m,n | a,i       | a,i,n     | a,i             | a,b,f,i              | a,i,k |
| Pb | a,b,c,f,i     | i         | b,i       |                 | a                    | b,c,f,i |
| Sn | a,b,d,e,f,n   |           | a,c,f,h   | a               | a.f.h.i              |     |
| Cd | p             |           |           |                 |                      |     |
| As | e,i           |           |           | a               | a,f,g                |     |
| Sb | i             |           |           |                 |                      |     |
| Se | f,g,          |           |           |                 | m                    |     |
| Ge | i,j           |           |           |                 |                      |     |
| Mn | a,f           |           |           |                 |                      | b   |
| Cu | f             |           |           |                 |                      |     |
| Ni |               |           |           |                 |                      | o   |
| Si | a,k           |           |           |                 |                      |     |

| | | | |
|---|---|---|---|
| a | GLC with conventional detectors | i | Direct AAS |
| b | GLC with AAS detector | j | Hydride generation AAS |
| c | GLC with ICPAES detector | k | ICPAES |
| d | GLC with mass spectrometric detector | l | X-ray fluorescence spectroscopy |
| e | Hydride generation — GLC | m | Neutron activation analysis |
| f | Conventional HPLC | n | Nuclear magnetic resonance spectroscopy |
| g | HPLC with ICPAES detector | o | Chemical ionisation mass spectrometry |
| h | Supercritical fluid chromatography | p | Anodic scanning volatimmetry |

# CHAPTER 1
# Introduction

A surprisingly large number of organometallic compounds occur in the environment. In this context the environment means inland waterways, potable water supplies, the oceans and also sedimentary matter, vegetation and animal life in inland waterways and the oceans.

Pollutants can enter river waters, coastal waters and open deep sea water. The concentrations in rivers and the coast would be expected to be higher than in the open sea due to the diluting effect of the latter. Having entered the water, organometallic pollutants will be partially absorbed onto bottom sediments and will enter the tissues of sea creatures such as fish, crustacea and also water plants such as algae and phytoplankton.

Some of the more volatile organometallic compounds are found in plant material, crops and biological materials such as fish, animal and human tissues and body fluids. Thus, organometallic compounds occur widely throughout the environment principally as compounds of mercury, tin, lead, arsenic and, to a lesser extent as compounds of germanium, antimony, copper, silicon, selenium, manganese and nickel. The distribution of these compounds throughout the environment—air, water, food—has toxicological implications which are of concern from the point of view of the health of humans, animals, fish, insects and birds, who, in one way or another, are all subject to contamination by organometallic compounds and who are all part of a food chain.

Considerations, other than toxicological, are involved in a consideration of the occurrence of organometallic compounds in the environment. These include the present and future roles of organometallic compounds in the biotechnologies that open industrial and regulatory options for bioremediation of unwanted toxic metal releases, recovery of precious metals, controls over bioactive and essential metals for agriculture and horticulture, biomining, and bioseparations of key industrial waste species subject to biotransformations by either endo- or exocellular organylation processes.

In all these areas, there is a growing recognition that high analytical sensitivity, alone, is insufficient. There is also a requirement from a detailed insight into the speciation of organometallic compounds occurring in the environment in order that a complete understanding can be obtained in processes occurring in nature. An example of such quantitative structure is the activity relationships illustrated, for example, by considering in the case of lead and tin, the effect of structure on human and mammalian toxicity.

Organometallic compounds enter the environment by three main routes. The first path is represented by the organometallic compounds produced by man such as those of lead and mercury. Lead enters chiefly as alkyl-lead compounds, used as an additive in gasoline which enters the air via gasoline spillages and possibly, to some extent, from automobile exhausts and then contaminates waterways and consequently river sediments and fish and plant life and also enters crops and animals in the fields. Similar comments can be applied to organomanganese compounds which, to some extent, are displacing lead as a petroleum additive and to organotin compounds which enter the environment, as for example, in antimoluscicide paints used on the hulls of ships and in harbour works.

The second source of organometallic compounds concerns inorganic substances such as industrial effluents and sewage introduced into the environment by man, and which are subsequently converted in nature to organometallic compounds. Classic cases of this are the inorganic mercury compounds and arsenic compounds, which in the environment or in animal tissues can be converted by bacteria to methylmercury and methylarsenic compounds. Probably the chief source of organomercury contamination is inorganic mercury entering rivers as an effluent from industries such as chloralkali works and industries which use alkalies in large quantities such as paper-making. It is believed that tin entering the environment via industrial effluents or mining operations as inorganic tin can be similarly converted in nature to organotin compounds. Certainly, we are only on the periphery of understanding what biotransformations occur when metal contaminated sewage sludge is disposed of as a fertiliser to agricultural land or to the oceans.

The third group of organometallic compounds which occur in the environment are those that are produced in waterways from naturally occurring metals. These include methylarsenic compounds produced by methylation of inorganic arsenic in fish or on sediment deposits, and probably include other elements such as antimony and selenium.

The list of organometallic compounds found in trace amounts in the environment has increased dramatically over the past few years and has necessitated the development of analytical methodology both for the purposes of first identifying new compounds and, secondly, of monitoring the concentrations of such compounds so that trends can be followed and working hypotheses developed. The occurrence of organometallic compounds in the environment is now systematically discussed.

The most powerful tools that are emerging to satisfy the needs of sensitivity and speciation are various combinations of chromatography, coupled with element specific detectors such as atomic absorption spectrometry or mass spectrometry, linked at the outlet end of the chromatography. These techniques have the advantages of being extremely sensitive, of providing a separation of sample components to meet the speciation needs and of confirming the metal and organic structure present by the specificity of the technique. Under these conditions nanogram quantities of each component can be identified and determined and, indeed, if sample preconcentration techniques are available, then subnanogram analysis becomes feasible.

The automation of techniques as exemplified by continuous lead in air monitoring programmes and the provision of on-line analysers in the biotechnology industries are areas in which rapid future developments are expected.

Before starting a discussion of the analytical procedures, the limited knowledge available to date concerning origins and toxic effects of organometallic compounds and the concentrations at which they occur in the environment is discussed below.

# CHAPTER 2
# Concentrations of Organometallic Compounds in the Ecosystem

## 2.1 THE RIVER ECOSYSTEM

The concentrations of organometallic compounds, and also the corresponding inorganic metals, found in various studies in river waters and sediments is summarised in Table 2.1. The higher concentrations of inorganic metal ions in sediments compared to those in river waters are in many cases self-evident. Also included in Table 2.1 are levels of inorganic metal ions known or suspected to occur as organometallic compounds in the environment but for which no organometallic data yet exists.

**Table 2.1.** Concentrations of organometallic compounds and inorganic metal ions in river waters and sediments (references in parentheses).

| | Organometallic compounds | | | Inorganic metals | |
|---|---|---|---|---|---|
| Compound | In river water ($\mu g\ L^{-1}$) | In sediment ($\mu g\ kg^{-1}$) | Metal | In river water ($\mu g\ L^{-1}$) | In sediment ($\mu g\ kg^{-1}$) |
| Tin | | | | | |
| Monomethyl | 0.001–0.04 (1–3,6) | — | Tin | 0.005–0.57 (1) | <0.01 |
| Dimethyl | 0.0007–0.005 (1–3,6) | — | | | |
| Trimethyl | 0.0006–0.005 (1–3,6) | — | | | |
| Dicyclohexyl | — | 10 (2) | | — | — |
| Tricyclohexyl | — | 75 (3) | | — | — |
| Monobutyl | 0.035–0.05 (3) | 280 (3) | | — | — |
| Dibutyl | 0.01–0.04 (3) | 140 (3) | | — | — |
| Tributyl | 0.005–0.015 (3) | 55 (3) | | — | — |
| Methyl mercury | 0.006–1.15 (4,5) | <0.01 0.02 (7) 0.2–0.4 (8) | Mercury | 0.009–13.0 | 910–46 800 |
| Alkyl lead | 50–530* (6) | <0.01 | Lead | 0.13–60 | 110–5 060 000 |
| Alkyl arsenic | — | — | Arsenic | 0.42–490 | 220–68 000 |
| Antimony | — | — | Antimony | 0.08–4.2 | 10–2900 |
| Germanium | — | — | Germanium | — | — |
| Selenium | — | — | Selenium | <0.0002–750 | 30–1000 |
| Manganese | — | — | Manganese | — | 340–9 640 000 |
| Copper | — | — | Copper | 0.11–200 | 70–244 000 |
| Nickel | — | — | Nickel | 1.5–4.5 | 1000–238 00 |

*530 $\mu g\ L^{-1}$ gasolene spillage.

Total metal tin compounds occur between 0.0023 and 0.05 µg L$^{-1}$ in river waters (0.005–0.57 µg L$^{-1}$ total methyltin plus inorganic tin). Total butyl tin compounds lie between 0.05 and 0.105 µg L$^{-1}$ in river water[11-17] and 470 µg kg$^{-1}$ in river sediments. The high levels of methyl mercury in the polluted River Waal (0.006–1.15 µg L$^{-1}$) and the high levels of alkyl lead originating from a gasoline spillage are notable (50–530 µg L$^{-1}$).

Butyl and cyclohexyl tin compounds and mono-, di- and tri-alkyl tin compounds[42-48] have been found in river and lake sediments. These probably originate from the use of organotin antifoulants on boats and pier works.

Concentrations of tributyltin and its degradation products have been determined in the surface microlayer of water at 74 locations in Ontario, Quebec and New York State[50]. The concentrations of tributyltin ranged from 1.9 to 473 µg L$^{-1}$ and were higher than concentrations previously reported for subsurface water. At 6 locations the concentration in the surface microsurface layer exceeded the 24 h LC$_{50}$ value of 1.3 µg L$^{-1}$ for adult rainbow trout. The most contaminated area was the mouth of the Moira River at Belleville Ontario, where the amount of tributyltin in the microlayer was 71% of the amount in the whole depth of the subsurface water. Higher concentrations of dibutyltin, monobutyltin and inorganic tin were also recorded in the surface microlayer.

Andren and Harris[7] have reported a methyl mercury concentration of 0.02 µg kg$^{-1}$ in unpolluted sediments, whilst Matsunaga and Takahaishi[8] found 0.2–0.4 µg kg$^{-1}$ methyl mercury in polluted sediments.

0.009 µg L$^{-1}$ methyl mercury has been found in European rainwater[5]. 0.006 µg L$^{-1}$ trimethyl tin[10] and < 0.001 µg L$^{-1}$ dibutyl tin and tributyl tin have been found in rainwater[3].

## 2.2 COASTAL WATER ECOSYSTEM

Due to direct coastal discharges of industrial waste and sewage and the flow of polluted rivers into estuaries and then the sea, the pollutant level in coastal waters would be expected to be higher than that in open seawater well away from the coasts. Available data on the concentrations of organometallic compounds and the corresponding inorganic metal ions in coastal waters and sediments are compared with levels in river waters in Table 2.2.

A number of observations can be made on this data. Concentrations of organotin compounds in river waters (expressed as µg L$^{-1}$) are lower than occur in river water sediments (expressed as µg kg$^{-1}$). Thus, in river waters, the total concentration of organotin compounds is 0.0023 to 0.105 µg L$^{-1}$, whereas in river sediments it is up to 470 µg kg$^{-1}$, i.e. a concentration factor (µg kg in sediment)/(µg L$^{-1}$ in water) in the sediment of 470/0.0023 to 470/0.105, i.e. 204 350 to 4476.

It is noteworthy that the range of concentrations of organotin in coastal waters (< 0.00001 to 3.28 µg L$^{-1}$ total mono-, di- and tri-alkyltin) is up to 30 times

**Table 2.2.** Concentrations of organometallic compounds and inorganic metal ions in rivers and coastal waters and sediments (references in parentheses).

| | In Rivers | | | | In Coastal Waters | | | |
|---|---|---|---|---|---|---|---|---|
| | Organometallic compound | | Total inorganic metal | | Organometallic compound | | Total Inorganic Metal | |
| | In water ($\mu g\ L^{-1}$) | In sediment ($\mu g\ kg^{-1}$) | In water ($\mu g\ L^{-1}$) | In sediment ($\mu g\ kg^{-1}$) | In water ($\mu g\ L^{-1}$) | In sediment ($\mu g\ kg^{-1}$) | In water ($\mu g\ L^{-1}$) | In sediment ($\mu g\ kg^{-1}$) |
| **Tin** | | | | | | | | |
| Monomethyl | 0.001–0.04 (1–3,6) | — | 0.005–0.57 (1) | <0.01 | — | — | <0.00001 | 1000–20,000 |
| Dimethyl | 0.0007–0.005 (1–3,6) | — | " | " | — | — | <0.00001 | " |
| Trimethyl | 0.0006–0.005 (1–3,6) | — | " | " | — | — | <0.00001 | " |
| Monoalkyl | — | — | — | — | <0.00001–1.22 | <0.01–1 (9,10) | <0.00001 | — |
| Dialkyl | — | — | — | — | 0.00074–0.46 | <0.01–0.1 (9,10) | <0.00001 | — |
| Trialkyl | — | — | — | — | <0.00001–1.6 | <0.01–0.01 (9,10) | <0.00001 | — |
| Monobutyl | 0.035–0.050 | 280 | — | — | — | — | — | — |
| Dibutyl | 0.010–0.040 | 140 | — | — | — | — | — | — |
| Tributyl | 0.005–0.015 | 55 | — | — | — | — | — | — |
| Alkyl mercury | 0.006–1.5 (4,5) | <0.01 | 0.009–13.0 | 910–46800 | 0.06 | <0.01 | 0.002–10.0 | <100–46800 |
| Alkyl arsenic | — | — | 0.42–490 | 220–68000 | 2.5–2.6 | <0.01 | 1.0–1.04 | 1600–11700 |
| Alkyl lead | 50–530 *(6) | <0.01 | 0.13–60 | 110–5060000 | — | <0.01 | 0.02–200 | 23000–38200 |
| Antimony | — | — | 0.08–0.42 | 10–2900 | — | — | 0.30–0.82 | 6200–13400 |
| Germanium | — | — | — | — | — | — | — | — |
| Selenium | — | — | <0.0002–750 | 30–1000 | — | — | <0.01–0.08 | 1500–9000 |
| Manganese | — | — | — | 340–9640000 | — | — | 0.35–250 | 21800–750000 |
| Copper | — | — | 0.11–200 | 70–244000 | — | — | 0.069–9.7 | 5400–84800 |
| Nickel | — | — | 1.5–4.5 | 1000–238000 | — | — | 0.2–15 | 30000–57000 |

*530 $\mu g\ L^{-1}$ gasolene spillage.

higher than in river waters (0.0023 to 0.105 µg L$^{-1}$ total methyl plus butyl tin compounds). This is presumably due to the fact that organotin compounds are more frequently used in the marine environment (harbour works and antifoulant treatment of boat hulls) than they are inland.

Concentrations of mono-, di- and tri-alkyl tin compounds found in Baltimore Harbour sediments averaged 8, 0.3 and 0.3 µg kg$^{-1}$ dry weight sediment whilst sediments taken in a relatively unpolluted area had much lower organotin content, 1.0, 0.1 and 0.01 µg kg$^{-1}$ for mono-, di- and tri-alkyl tin[9,10].

In the case of organolead compounds the concentrations of these in river waters is higher than in sediments. Thus in river waters the total concentration of organolead compounds is 50–530 µg L$^{-1}$, whilst in the sediment it is < 0.01 µg kg$^{-1}$ (<0.00001 mg kg$^{-1}$) i.e. a concentration factor in the sediment (µg kg$^{-1}$ in sediment)/(µg L$^{-1}$ in water) of 0.01/50 to 0.01/530 = 0.0002 to 0.00002.

In the case of organomercury compounds the concentration of organomercury compounds in river water is usually higher than in sediments. Thus, in river waters the total concentration of organomercury compounds is 0.006–1.5 µg L$^{-1}$, whilst in the sediment it is < 0.01 µg L$^{-1}$ (< 0.00001 µg kg$^{-1}$), i.e. a concentration factor (µg kg$^{-1}$ in sediment)/(µg L$^{-1}$ in water) of 0.01/0.006 to 0.01/1.5 = 1.7 to 0.007.

In coastal waters the total concentration of organomercury compounds is 0.06 µg L$^{-1}$, whilst in the sediment it is < 0.01 µg kg$^{-1}$ (< 0.00001 mg kg$^{-1}$) i.e. a concentration factor (µg kg$^{-1}$ in sediment)/(µg L$^{-1}$ in water) of 0.01/0.06 = 0.17. At the higher end of its concentration range (1.5 µg L$^{-1}$) the concentration of organomercury in river waters is considerably higher than that in coastal waters (0.06 µg L$^{-1}$), presumably because the mercury originates inland as industrial effluents and agricultural discharges and proceeds via rivers and coastal discharges to coastal waters.

In the case of organoarsenic compounds a similar situation would be assumed to prevail. The concentration of organoarsenic in coastal waters (2.5–2.6 µg L$^{-1}$) is considerably higher than the 0.1 µg kg$^{-1}$ level (0.00001 mg kg$^{-1}$) occurring in sediments, i.e. a concentration factor of (µg kg$^{-1}$ in sediment)/(µg L$^{-1}$ in water) of 0.01/2.5 = 0.004.

Of the organometallic compounds discussed above tin is the exception in that it concentrates from water into sediments with a maximum observed concentration factor (µg kg$^{-1}$ tin in sediment)/(µg L$^{-1}$ tin in water) being between approximately 5000 and 200 000 (see Table 2.3).

Several governments (UK and France and several states in the USA) have banned the use of organotin compounds in recreational craft whilst the EU and Scandinavian countries are debating this issue[26]. Langsten et al.[18] have carried out a detailed study of the occurrence of organotin compounds in water, sediments and benthic organisms in Poole Harbour, Dorset, UK.

Poole Harbour is a series of interconnecting basins utilised by over 7000 leisure craft and small numbers of larger commercial vessels. The narrow harbour

# CONCENTRATIONS OF ORGANOMETALLIC COMPOUNDS IN THE ECOSYSTEM

**Table 2.3.** Concentration factors* of organometallic and ionic metal levels in rivers and coastal waters.

| Element | Rivers | | Coastal waters | |
|---|---|---|---|---|
| | Organometallic | Ionic metal | Organometallic | Ionic metal |
| Tin | $4.5 \times 10^3 - 2 \times 10^5$ | $0.017-2$ | — | $10^8 - 2 \times 10^9$ |
| Lead | $0.00002-0.0002$ | $8.5 \times 10^2 - 8.4 \times 10^4$ | — | $1.9 \times 10^2 - 1.5 \times 10^6$ |
| Mercury | $0.007-1.7$ | $3.6 \times 10^3 - 10^5$ | $0.17$ | $4.7 \times 10^3 - 5 \times 10^4$ |
| Arsenic | — | $1.4 \times 10^2 - 5.2 \times 10^2$ | $0.004$ | $1.6 \times 10^3 - 1.17 \times 10^4$ |
| Antimony | — | $1.2 \times 10^2 - 6.9 \times 10^3$ | — | $1.6 \times 10^4 - 2.1 \times 10^4$ |
| Selenium | — | $1.3-1.5 \times 10^5$ | — | $1.1 \times 10^5 - 1.5 \times 10^5$ |
| Manganese | — | $3.4 \times 10^7 - 10^{12}$ | — | $3 \times 10^3 - 0.6 \times 10^6$ |
| Copper | — | $6.4 \times 10^2 - 1.2 \times 10^3$ | — | $8.7 \times 10^4 - 7.7 \times 10^4$ |
| Nickel | — | $6.7 \times 10^2 - 5.3 \times 10^4$ | — | $3.8 \times 10^3 - 1.5 \times 10^5$ |

*$\mu g\ kg^{-1}$ substance in sediment, (analytical data from Table 2.1)
$\mu g\ L^{-1}$, substance in water.

entrance and limited tidal range resulted in poor flushing. A survey of 19 sites around the harbour between July 1985 and February 1987 showed that concentrations or tributyltin in water ranged from 2 to 646 ng tin $L^{-1}$, with the highest levels (234–646 ng tin $L^{-1}$) occurring within the marinas. Seasonal variations were related to boat traffic. Sedimentary tributyltin ranged from 0.02 ng tin per g in the harbour mouth to 0.052 µg tin per g near marinas and moorrings. High organotin concentrations were also detected in polychaetes, snails and bivalves. Tributyltin concentrations in much of Poole Harbour exceeded the Environmental Quality Targets for the protection of marine organisms (20 ng $L^{-1}$ as tributyltin, 8 ng $L^{-1}$ as tin) and may account for the poor recruitment of bivalves at heavily contaminated sites.

Organocompounds of lead, mercury and arsenic concentrate from water into sediments to a much smaller extent with a maximum concentration factor of about 2 for mercury in rivers (Table 2.3).

The corresponding inorganic metal ions, on the other hand, concentrate into sediments to a much greater extent. Thus, maximum observed concentration factors for inorganic tin, lead, mercury and arsenic are, respectively $2 \times 10^9$, $1.15 \times 10^6$, $10^5$ and $1.17 \times 10^4$. Even at the lower ends of the range, minimum observed concentration factors for inorganic lead, mercury and arsenic at 190, 3600 and 140 are considerably higher than those observed for the corresponding organometallic compounds. Tin is an exception in that the minimum concentration factor for inorganic tin (0.02) is lower than the factor for organic tin ($5 \times 10^3 - 2 \times 10^5$). Data is not available for concentration factors between water and sediments of organometallic compounds of other elements of environmental concern. It is seen in Table 2.3, however, that the maximum observed concentration factors for the inorganic forms of these elements are generally in the range of $10^2$ to $10^6$ for both river waters and coastal waters, i.e. these elements in their inorganic form show a strong tendency to concentrate from the water column into sediments. These observations regarding the transfer of inorganic forms of elements from the water column to sediments are supported by data available from the North Sea as discussed in Section 2.3

The wide ranges of concentration factors quoted in Table 2.3 are a reflection of the variability of data reported in various sources and these variations are no doubt, in part, due to differences in water chemistry (e.g. pH and salinity) from sample to sample.

## 2.3 OPEN SEAWATER ECOSYSTEM

The determination of trace metals and organometallic compounds in open seawater stretches the abilities of the analytical chemist and oceanographer and their equipment to the limit.

Metal concentrations are exceedingly low, often in the nanogram per litre (ng $L^{-1}$) range and contamination during sampling and analysis present many problems. So much so, in fact, that any data collected prior to 1975 can be disregarded as suspect. Only post-1975 data is quoted here. The values now being obtained for many elements are considerably lower than those obtained prior to 1975 as illustrated in Table 2.4

Concentrations of organometallic compounds found in seawater are tabulated in Table 2.5. Nanogram $L^{-1}$ levels of mono-, di- and tri-alkyllead compounds occur in seawater[42-48].

Traces of organotin compounds are only found in certain coastal areas where these compounds are used as antifoulants on boats or harbour works.

This distribution of inorganic elements and, consequently, the potential for the occurrence of the corresponding organometallic compounds, in various oceans varies considerably (Table 2.6). Thus, in the cases of copper and nickel in the Pacific, concentrations are between 5 and 50 times consensus values for seawater. In Tables 2.7 and 2.8, respectively, are compared concentrations of organometallic compounds and inorganic metal ions found in rivers, coastal waters and open seawaters, and the corresponding sediments. As would be expected inorganic metal concentrations decrease progressively from rivers to the open sea. This also occurs in the case of organomercury and organolead compounds. Maximum organoarsenic concentrations are similar (2.5 μg $L^{-1}$) in

**Table 2.4.** Reduction in reported values for elements in open seawater due to control of contamination.

| Element | Reported prior to 1975 (μg $L^{-1}$) | Reported post 1980 (μg $L^{-1}$) | Consensus value 1990 (μg $L^{-1}$) |
|---|---|---|---|
| Copper | 0.5–6 | 0.1–0.2 | 0.05 |
| Nickel | 2 | 0.25–0.39 | 0.17 |

*Consensus value defined as level at which pollution is deemed not to have occurred.

Table 2.5. Organometallic compounds in seawater.

| | Organometallic compound ($\mu g\ L^{-1}$) | Reference |
|---|---|---|
| Mercury | | |
| Me Hg in seawater | 0.06 | 19 |
| Arsenic | | |
| Irish Sea | 2.49–2.65 | 20 |
| Tin | | |
| Gulf of Mexico | | 21 |
| SnIV | 0.0022–0.062 | |
| MeSn | <0.00001–0.015 | |
| $Me_2Sn$ | 0.00074–0.007 | |
| $Me_3Sn$ | <0.00001–0.00098 | |
| Total Sn | 0.0036–0.085 | |
| Old Tampa Bay | | 21 |
| SnIV | <0.0003–0.0027 | |
| MeSn | 0.00086–0.0011 | |
| $Me_2Sn$ | 0.0006–0.002 | |
| $Me_3Sn$ | <0.00001–0.00095 | |
| Total Sn | 0.0025–0.005 | |
| Estuary | | 21 |
| SnIV | 0.0003–0.020 | |
| MeSn | <0.00001–0.008 | |
| $Me_2Sn$ | 0.00079–0.0022 | |
| $Me_3Sn$ | <0.00001–0.0011 | |
| Total Sn | 0.0025–0.023 | |
| Harbour Water | | 22 |
| $Me_2Sn$ | <0.01–0.02 | |
| $Me_3Sn$ | <0.01–0.02 | |
| $SnH_4$ | 0.2–20 | |
| $Me_4Sn$ | <0.01–0.3 | |
| $BuSnH_3$ | 0.05–0.3 | |
| Surface Water | | 23 |
| SnIV | 0.001–0.009 | |
| $BuSnH_3$ | 0.01–0.06 | |
| $Bu_2SnH_2$ | 0.13–0.46 | |
| $Bu_3SnH_2$ | 0.06–0.78 | |
| Bottom Water | | 23 |
| SnIV | 0.003–0.005 | |
| $BuSnH_3$ | 0.03–0.04 | |
| $Bu_2SnH_2$ | 0.13 | |
| $Bu_3SnH$ | 0.01–0.10 | |
| Estuary Water | | |
| $Bu_3Sn$ | 0.08–0.19 | |
| Bay Samples | | 21 |
| SnIV | 0.003–0.02 | |
| MeSn | 0.0007–0.008 | |

*continued overleaf*

**Table 2.5.** (*continued*)

| | Organometallic compound (µg L$^{-1}$) | Reference |
|---|---|---|
| **Tin** | | |
| Me$_2$Sn | 0.0008–0.002 | |
| Me$_3$Sn | 0.0003–0.001 | |
| Total Sn | 0.0002–0.023 | |
| Lake Michigan | | 21 |
| Adjacent to coast | | |
| SnIV | 0.08–0.49 | |
| MeSnCl$_3$ | 0.006–0.0013 | |
| Me$_2$SnCl$_2$ | <0.0001–0.063 | |
| BuSnCl$_3$ | 0.002–1.22 | |
| Bu$_2$SnCl$_2$ | 0.01–1.6 | |
| San Diego Bay | | 21 |
| Surface Water | | |
| SnIV | 0.006–0.038 | |
| MeSnCl$_3$ | 0.0002–0.0008 | |
| Me$_2$SnCl$_2$ | 0.015–0.045 | |
| BuSnCl$_3$ | <0.0001 | |
| Bu$_2$SnCl$_2$ | <0.0001 | |
| San Francisco Bay | | 21 |
| SnIV | 0.0002–0.0003 | |
| MeSnCl$_3$ | <0.0001 | |
| Me$_2$SnCl$_2$ | <0.0001 | |
| BuSnCl$_3$ | <0.0001 | |
| Bu$_2$SnCl$_2$ | <0.0001 | |
| Coastal Adjacent to | | 21 |
| San Francisco | | |
| SnIV | 0.0003–0.0008 | |
| MeSnCl$_3$ | <0.0001 | |
| Me$_2$SnCl$_2$ | <0.0001 | |
| BnSnCl$_3$ | <0.0001 | |

coastal and seawaters, whilst the maximum observed concentration of organotin compounds in seawater can be considerably greater (2.7 µg L$^{-1}$) than in rivers (0.1 µg L$^{-1}$ maximum).

Some detailed data is available on the pollution load in the North Sea. Data in Table 2.9 shows that the lead elements potentially capable of forming organometallic compounds in biosynthesis originates principally from rivers and dredging spoils with a not inconsiderable proportion originating as atmospheric pollution from factory stack discharges. Direct coastal discharges and dumping at sea account for only a small proportion of the total pollution load on the North Sea.

The following assumptions are made:

Concentration element in Sediment $A$ µg kg$^{-1}$
Concentration element in water $B$ µg L$^{-1}$

Table 2.6. Inorganic metal contents of open seas (µg L$^{-1}$).

| | Pacific | Atlantic | Norwegian Sea | Sargossa Sea | Arctic Sea | Unidentified seas | Overall range | Consensus value | Coastal waters |
|---|---|---|---|---|---|---|---|---|---|
| Copper | 0.3–2.8 | — | 0.08–0.10 | 0.072–0.081 | 0.097 | 0.0063–2.8 | 0.072–2.8 | 0.05 | 0.069–9.7 |
| Nickel | 0.15–0.93 | — | 0.17–0.20 | 0.26–0.27 | 0.099 | 0.15–0.93 | 0.099–0.93 | 0.17 | 0.2–15 |
| Lead | <0.01–0.8 | 0.00017–0.003 | 0.025–0.065 | 0.00004 | 0.01–0.021 | 0.000041–9.0 | 0.00004–9.0 | a | 0.2–200 |
| Mercury | — | 0.021–0.078 | — | — | — | 0.002–0.078 | 0.002–0.078 | <0.2 | 0.002–10 |
| Selenium | — | — | — | — | — | 0.00095–0.029 | 0.00095–0.029 | a | <0.01–0.08 |
| Tin | — | — | — | — | — | 0.02–0.05 | 0.02–0.05 | a | — |
| Manganese | — | — | — | — | — | 0.018 | 0.018 | a | 0.35–250 |

[a]Consensus values not designated.

**Table 2.7.** Comparison of organometallic concentrations in rivers, coastal waters, open seas and sediments (references in parentheses).

| | Rivers | | Coastal Waters | | Open Seas | |
|---|---|---|---|---|---|---|
| | Water ($\mu g\ L^{-1}$) | Sediment ($\mu g\ kg^{-1}$) | Water ($\mu g\ L^{-1}$) | Sediment ($\mu g\ kg^{-1}$) | Water ($\mu g\ L^{-1}$) | Sediment ($\mu g\ kg^{-1}$) |
| Tin | | | | | | |
| Monomethyl | 0.001–0.04 (1–3,6) | — | — | — | <0.00001–0.015 (1,22–24) | — |
| Dimethyl | 0.0007–0.005 (1–3,6) | — | — | — | <0.00001–0.063 (1,22–24) | — |
| Trimethyl | 0.0006–0.005 (1–3,6) | — | — | — | <0.00001–0.02 (1,22–24) | — |
| Dicyclohexyl | — | 10 | — | — | — | — |
| Tricyclohexyl | — | 75 | — | — | — | — |
| Monobutyl | 0.035–0.050 | 280 | — | — | <0.0001–0.3 (1,22–24) | — |
| Dibutyl | 0.010–0.040 | 140 | — | — | <0.0001–1.6 (1,22–24) | — |
| Tributyl | 0.005–0.015 | 55 | — | — | 0.06–0.78 (1,22–44) | — |
| Monoalkyl | — | — | <0.00001–1.22 | <0.01 | — | 1*–8** |
| Dialkyl | — | — | 0.00074–0.46 | <0.01 | — | 0.1*–1** |
| Trialkyl | — | — | <0.00001–1.6 | <0.01 | — | 0.01*–0.3** |
| Methylmercury | 0.006–1.15 (4,5) | — | 0.06 | — | 0.07 (19) | — |
| Alkyl lead | 50–530† (6) | — | <0.00001 | — | — | — |
| Alkyl arsenic | — | — | 2.5–2.6 | — | 2.49–2.63 (20) | — |

†Value of 530 $\mu g\ L^{-1}$ due to gasoline spillage in the area.
*Unpolluted waters.
**Polluted waters.

# CONCENTRATIONS OF ORGANOMETALLIC COMPOUNDS IN THE ECOSYSTEM

**Table 2.8.** Comparison of total inorganic metal ion concentrations in rivers, coastal waters, open seas and sediments (references in parentheses).

|  | Rivers | | Coastal Waters | | Open Seas* | |
|---|---|---|---|---|---|---|
|  | Water ($\mu g\ L^{-1}$) | Sediment ($\mu g\ kg^{-1}$) | Water ($\mu g\ L^{-1}$) | Sediment ($\mu g\ kg^{-1}$) | Water ($\mu g\ L^{-1}$) | Sediment ($\mu g\ kg^{-1}$) |
| Tin | 0.004–0.57 (1) | <0.01 | — | 1000–20 000 | 0.02–0.05 | 1000–20 000 (26) |
| Mercury | 0.009–13.0 | 910–46 800 | 0.002–10.0 | <100–46 800 | 0.002–0.078 | <100–46 800 (25) |
| Arsenic | 0.42–490 | 220–68 000 | 1.0–1.04 | 1600–11 700 | 0.00004–9.0 | — |
| Lead | 0.13–60 | 110–5 060 000 | 0.02–200 | 23 000–38 200 | — | — |
| Antimony | 0.08–0.042 | 10–2900 | 0.30–0.82 | 6200–13 400 | — | — |
| Germanium | — | — | — | — | — | — |
| Selenium | <0.0002–750 | 30–1000 | <0.001–0.08 | 1500–9000 | 0.00095–0.029 | — |
| Manganese | — | 340–9 640 000 | 0.35–250 | 218 000–750 000 | 0.018 | — |
| Copper | 0.11–200 | 70–244 000 | 0.069–9.7 | 5400–84 800 | 0.072–2.8 | — |
| Nickel | 1.5–4.5 | 1000–238 000 | 0.2–15.0 | 30 000–57 000 | 0.099–0.93 | — |

*Includes Pacific, Atlantic, Norwegian, Adriatic, Saragossa and Japanese Seas, etc. Concensus values in sea water, mercury < 0.2 μg $L^{-1}$, Copper 0.05 μg $L^{-1}$ and Nickel 0.17 μg $L^{-1}$.

Table 2.9. Pollution load on North Sea (tonne per annum).

| Element | From all sources | From river pollution | From coastal discharges | From sea dumping | From dredging spills | From atmosphere pollution |
|---|---|---|---|---|---|---|
| Arsenic | 1030 | 554 | 206 | 5 | 5 | 230 |
| Lead | 9441 | 2554 | 150 | 377 | 3440 | 2920 |
| Mercury | 117.5 | 27 | 7 | 2.5 | 30 | 51 |
| Copper | 8647 | 2600 | 276 | 360 | 1469 | 3942 |
| Nickel | 5621 | 2466 | 500 | 97 | 989 | 1569 |
| Total | 24 856.5 | 8201 | 1139 | 836.5 | 5933 | 8712 |

Concentration factor $A/B = F$
Volume of water in North Sea water column $V_w$ L (assumed $10^{15}$ L)
Weight of sediment in North Sea water column $W_s$ kg (assumed $10^{11}$ kg)

The discharges into the North Sea consist of water soluble and water insoluble forms of the elements. The insoluble forms will fall as sediment to the ocean floor or be suspended in the water whilst the soluble forms might either remain in solution or absorb on to sedimentary matter in the sea over a period of time. In Table 2.10 are shown the total weight of soluble plus insoluble forms of five elements of environmental concern and capable of forming organometallic compounds in the North Sea at the start of 1984 (Column A), the additional amounts added as pollution load during 1984 (Column B), and the total amounts present at the end of 1984 (Column C).

An attempt is made below to calculate the distribution between the water phase and the sedimentary matter phases of the total 24 856 tonne pollution load introduced into the North Sea during 1984. These calculations are based on known concentration factors between the water and solid phases (Table 2.3).

Total weight of element in water column (i.e. in water plus sediment)

$$= M \text{ tonne} = 10^{12} M \text{ μg}$$

Table 2.10. Weight (tonne) of soluble plus insoluble forms of elements present in the north sea.

| Element | A<br>Present at<br>Start of 1984 | B<br>Added<br>During 1984 | C<br>Present at<br>End of 1984 |
|---|---|---|---|
| Arsenic | 1390 | 1030 | 2420 |
| Lead | 12 792 | 9441 | 22 233 |
| Mercury | 159.2 | 117.5 | 276.7 |
| Copper | 11 717 | 8647 | 20 364 |
| Nickel | 7616 | 5621 | 13 237 |
| Total | 33 680 | 24 856 | 58 536 |

The weight of the element in $W_s$ kg sediment plus $V_w$ L water is then $W_sA + V_wB$ µg, i.e. $M \times 10^{12} = W_sA + V_wB$.
Substituting the above values for $W_s$ and $V_w$

$$A = \frac{M \times 10^{12} - 10^{15}B}{10^{11}}$$

Solving for $A$ and $B$:

$$\text{Concentration of element in water} = B = \frac{10^{12}M}{10^{15}\left(1 + \dfrac{F}{10.000}\right)}$$

$$= \frac{M}{10^3\left(1 + \dfrac{F}{10.000}\right)} \text{ µg L}^{-1} \text{ element in water} \quad (1)$$

$$\text{Concentration of element in sediment} = A = \frac{10^{12}MF}{10^{15}\left(1 + \dfrac{F}{10.000}\right)}$$

$$= \frac{MF}{10^3\left(1 + \dfrac{F}{10.000}\right)} \text{ µg kg}^{-1} \text{ element in sediment} \quad (2)$$

As an example, consider the case of organic and inorganic arsenic in coastal waters. The values used in calculations for (a) concentration factor ($F$) of each form of arsenic between water and sediment, (b) weight in tonnes of soluble pluse insoluble inorganic and organic arsenic in the water column at the start and the end of 1984 ($M$ and $M^1$ tonne respectively) and (c) the resulting concentrations of each form of arsenic in the water column and in the sediments at the start and end of 1984 is quoted in Table 2.11.

The resulting mass balance of inorganic and organic arsenic between water and sediment during 1984 is summarised in Table 2.12. The results in Table 2.12 show that the introduction of an additional 1030 tonne of inorganic arsenic pollution load on to the North Sea during 1984 leads to a much greater increase in arsenic content of the sediment (1440–5616 µg kg$^{-1}$ increase) than of the water (0.9–0.48 µg L$^{-1}$ increase). However, because the assumed volume of the water column in the North Sea ($10^{15}$ L) is much greater than the assumed weight of sediment present ($10^{11}$ kg), the actual weight of inorganic arsenic dissolved in the water (900–480 tonne arsenic) is greater that the weight of arsenic which become adsorbed on to the sediments (144–562 tonne arsenic).

In the case of organic arsenic the 100 tonne additional pollution load introduced during 1984 leads to a smaller increase in organoarsenic content of the sediment (0.0004 µg kg$^{-1}$ increase) than of the water (0.1 µg L$^{-1}$ increase). Almost all of the added 100 tonne of organic arsenic will remain dissolved in the water phase.

Table 2.11. Distribution of organoarsenic and inorganic arsenic in the North Sea during 1984.

| Concentration factor $F$ (see Table 2.3) | Weight (tonne) of soluble plus insoluble arsenic in water column (Table 2.10) at | | | Concentration of arsenic in water (μg L⁻¹) | | | Concentration of arsenic in sediment (μg kg⁻¹) | | |
|---|---|---|---|---|---|---|---|---|---|
| | Start 1984 | End 1984 | Added during 1984 | Start 1984 | End 1984 | Added during 1984 | Start 1984 | End 1984 | Added during 1984 |
| | $M$ | $M^1$ | $M^1 - M$ | $M^*$ $\dfrac{M^*}{10^3\left(1+\dfrac{F}{10^4}\right)}$ | $M^{1*}$ $\dfrac{M^{1*}}{10^3\left(1+\dfrac{F}{10^4}\right)}$ B | $M^1 - M^*$ $\dfrac{M^1 - M^*}{10^3\left(1+\dfrac{F}{10^4}\right)}$ | $MF^{**}$ $\dfrac{MF^{**}}{10^3\left(1+\dfrac{F}{10^4}\right)}$ | $M^1F^{**}$ $\dfrac{M^1F^{**}}{10^3\left(1+\dfrac{F}{10^4}\right)}$ A | $(M^1 - M)F^{**}$ $\dfrac{(M^1-M)F^{**}}{10^3\left(1+\dfrac{R}{10^4}\right)}$ |
| Inorganic arsenic | | | | | | | | | |
| 1600–11700 | 1390 | 2420 | 1030 | 1.20–0.64 | 2.09–1.11 B' | 0.90–0.48 | 1920–7,488 | 3344–12,987 A' | 1440–5,616 |
| Organic arsenic | | | | | | | | | |
| 0.004 | 200 | 300 | 100 | 0.2 | 0.3 | 0.1 | 0.0008 | 0.0012 | 0.0004 |

*From equation (1).
**from equation (2).

CONCENTRATIONS OF ORGANOMETALLIC COMPOUNDS IN THE ECOSYSTEM 19

Table 2.12. Mass balance of inorganic and organic arsenic between water and sediments during 1984.

| | Inorganic arsenic | | Organic arsenic | |
|---|---|---|---|---|
| | In water | In sediment | In water | In sediment |
| | Start of 1984 | | | |
| Conc. of arsenic (C) | 1.20–0.64 µg L$^{-1}$ | 1920–7488 µg kg$^{-1}$ | 0.2 µg L$^{-1}$ | 0.0008 µg kg$^{-1}$ |
| Weight (tonne) of arsenic in $10^{15}$ L water column ($10^3 \times C$) | 1200–640 tonne | — | 200 tonne | — |
| Weight (tonne) of arsenic in $10^{11}$ kg sediment (C)/10 | — | 192–749 tonne | — | 0.00008 tonne |
| | End of 1984 | | | |
| Conc. of arsenic (C) | 2.09–1.10 µg L$^{-1}$ | 3346–13 049 µg kg$^{-1}$ | 0.3 µg L$^{-1}$ | 0.0012 µg kg$^{-1}$ |
| Weight (tonne) of arsenic in $10^{15}$ L water column ($10^3 \times C$) | 2090–1100 tonne | — | 300 tonne | — |
| Weight (tonne) of arsenic in $10^{11}$ kg sediment (C)/10 | — | 335–1305 tonne | — | 0.00012 tonne |
| | Increase in Pollution Load During 1984 | | | |
| Conc. of arsenic (C) | 0.90–0.48 µg L$^{-1}$ | 1440–5616 µg kg$^{-1}$ | 0.1 µg L$^{-1}$ | 0.0004 µg kg$^{-1}$ |
| Weight (tonne) of arsenic in $10^{15}$ L water column ($10^3 \times C$) | 900–480 tonne | — | 100 tonne | — |
| Weight (tonne) of arsenic in $10^{11}$ kg sediment (C)/10 | — | 144–562 (tonne) | — | 0.00004 tonne |

| | Inorganic arsenic | Organic arsenic |
|---|---|---|
| Total weight (tonne) of arsenic in water plus sediment | 1044–1042 tonne | 100 tonne |
| Actual total weight (tonne) of arsenic added to sea in 1984 | 1030 tonne | 100 tonne |

In a further example we consider the case of inorganic and organic tin in water. In the case of tin, as has been previously discussed (see Table 2.3), adsorption behaviour on to sediments is very different to that occurring in the case of arsenic. As the concentration factor data in Table 2.3 shows, inorganic arsenic is strongly adsorbed on to sediments from water ($F = 1600–11\,700$), whereas organoarsenic is not adsorbed ($F = 0.017–2.0$) whereas organotin compounds are strongly adsorbed on to sediments ($F = 4.5 \times 10^3 – 2.4 \times 10^5$).

Unfortunately, no data is available on the total weight of soluble plus insoluble inorganic and organometallic tin in a given volume of the oceans. Consequently, as an example the following assumptions are made: 4 tonne of inorganic tin and 1 tonne of organotin are present in $10^{15}$ L of water containing $10^{11}$ kg of sediment. This data is summarised in Table 2.13.

The introduction of 4 tonne of inorganic tin leads to a greater increase in the tin content of the water (0.004 µg kg$^{-1}$) than of the sediment ($6.8 \times 10^{-5}$ to $8 \times 10^{-3}$ µg kg$^{-1}$). The situation is reversed, however, in the case of organotin where the increase in organotin content of the water ($6.9 \times 10^{-4} – 5.10 \times 10^{-5}$ µg L$^{-1}$) is much lower than that in the sediment (3.1–10.0 µg kg$^{-1}$) i.e. almost all of the organotin compound is adsorbed on to the sediment. In the case of inorganic tin almost all of the 4 tonne pollution load remains dissolved in the water whilst in the case of organotin, almost all of the 1 tonne pollution load adsorbs on to the sediment.

Studies such as those described above are of great value in establishing for any water system the distribution that occurs of organic and inorganic metal pollution between natural waters and sedimentary matter. It also infers that analysis of the water alone is inadequate for assessing the effects of pollution. Very frequently pollution is concentrated in sedimentary matter and it is equally essential to analyse this.

It is a known fact that the total soluble metal content of the North Sea is not increasing annually at the rate which the annual pollution load would infer and it is concluded that this is due to two mechanisms. Firstly, the additional soluble metal and organometallic compounds added each year are being dispersed to areas outside the North Sea and, secondly, that some of the added metals and organometallic compounds are being adsorbed on to sediments which fall to the sea bed, i.e. are converted to insoluble metals. Thus, the picture is not as bright as it might at first seem. Although the concentration of soluble metals in the water is not increasing at the expected rate, based on annual pollution inputs, these substances are still polluting the ecosystem by dispersal through the oceans and by contamination of sedimentary matter in the oceans.

## 2.4 FISH AND OTHER CREATURES IN THE ECOSYSTEM

The concentration of substances picked up from water by sea creatures such as fish and crustacea and by phytoplankton and plant matter is dependent upon the concentration of the substance in the water and, to some extent, upon its

# CONCENTRATIONS OF ORGANOMETALLIC COMPOUNDS IN THE ECOSYSTEM 21

**Table 2.13.** Distribution of organotin compounds and inorganic tin in water and sediments.

| Concentration factor $F$ (see Table 2.3) | Weight (tonne) of tin in water column ($10^{15}$ L) $M$ | Concentration of tin in water (µg L$^{-1}$) $B = \dfrac{M^*}{10^3 \left(1 + \dfrac{F}{10^4}\right)}$ | Weight of tin in $10^{15}$ L water ($B \times 10^3$) | Concentration of tin in sediment (µg kg$^{-1}$) $A = \dfrac{MF^{**}}{10^3 \left(1 + \dfrac{F}{10^4}\right)}$ | Weight of tin in $10^{11}$ kg sediment ($A/10$) | Weight of tin in $10^{15}$ L water plus $10^{11}$ kg sediment $B \times 10^3 + A/10$ |
|---|---|---|---|---|---|---|
| Inorganic tin 0.017–2.0 | 4 | 0.004 | 4 | $6.8 \times 10^{-5}$–$8 \times 10^{-3}$ | $6.8 \times 10^{-6}$–$8 \times 10^{-4}$ | 4 |
| Organic tin $4.5 \times 10^3$–$2 \times 10^5$ | 1 | $6.9 \times 10^{-4}$–$5 \times 10^{-5}$ | 0.69–0.05 | 3.1–10.0 | 0.31–1.0 | 1 |

*from equation (1).
**from equation (2).

concentration in sedimentary matter. Many creatures bioaccumulate toxicants from the water as discussed in Chapter 5 and as a consequence their concentration in the organism is many times higher than that in the water in which they live. Once the concentration in the organism exceeds a certain level then harmful effects or mortalities occur. Analysis of creatures is therefore a very useful means of ascertaining the cause of adverse effect or death in creatures. Much work has been carried out on the determination of toxicants in creatures and this is discussed below.

Many polluting organometallic compounds have been found in fish and crustacea. These include tetraalkyl lead compounds (up to 18.9 mg kg$^{-1}$ in carp), methyl mercury compounds (up to 11.5 mg kg$^{-1}$ in tuna) and alkylarsenic compounds (up to 23 mg kg$^{-1}$ in molluscs). Many of these substances are highly toxic and their effect on sea creatures and the humans who eat them will be discussed in later chapters.

### 2.4.1 FISH

Details of the occurrence in fish of organolead and organomercury compounds are given in Tables 2.14 and 2.15. These originate predominantly from the use of alkyl lead compounds in petroleum and the methylation of inorganic mercury released into the ecosystem as effluents in the chloroalkali process.

Concentrations of organometallic compounds and the corresponding total ionic metal found in fish are compared in Table 2.16. In the cases of lead and mercury, certainly, the organometallic content of the fish is an appreciable proportion of the total mercury content.

The organs of fish such as opercle, kidney and liver have the property of concentrating metals and organometallic compounds such that the concentration in the organ is appreciably higher than in the whole fish tissue. This is illustrated

**Table 2.14.** Concentrations (µg kg$^{-1}$) of organolead compounds found in environmental fish from Von Endt et al.[27] Reproduced by permission of American Chemical Society.

| Compound | Carp | Bass | Small mouth bass | Pike | White sucker | Range, all types |
|---|---|---|---|---|---|---|
| Me$_4$Pb | 140 | <1 | — | — | — | 140 |
| Me$_3$EtPb | <1 | — | — | — | — | <1 |
| Me$_2$Et$_2$Pb | 1430 | — | 57 | 100 | — | 57–1430 |
| MeEt$_3$Pb | 140 | — | 190–250 | 150–170 | 290 | 140–290 |
| Et$_4$Pb | 780–7500 | — | 1200–1830 | 1020–1120 | 2950–4380 | 780–7500 |
| Me$_3$Pb$^+$ | <10 | — | <10 | 200–210 | 90–200 | <10–210 |
| Me$_2$Pb$^{2+}$ | 360 | — | — | — | — | 360 |
| Et$_3$Pb$^+$ | 90–1210 | — | 220–660 | 53 | 2170–3430 | 54–3430 |
| Et$_2$Pb$^{2+}$ | 710–1310 | — | 90–2750 | — | 2200–4300 | 90–4300 |
| Pb$^{2+}$ | 1280–4130 | — | 250–300 | 1040–1190 | 3610–3480 | 250–4130 |
| Total excl Pb$^{2+}$ | 5090–18 940 | <1 | 1760–5555 | 1520–1650 | 7700–12 600 | 1630–16 660 |

# CONCENTRATIONS OF ORGANOMETALLIC COMPOUNDS IN THE ECOSYSTEM

**Table 2.15.** Concentrations ($\mu g\ L^{-1}$) of organomercury compounds found in environmental fish (references in parentheses).

| Compound | Whiting | Sardine | Turbot | Halibut | Coho Salmon | Salmon | Red Tuna | White Tuna | Pike | Trout |
|---|---|---|---|---|---|---|---|---|---|---|
| $MeHg^+$ | | | | | | | | 930 (32) | | |
| $EtHg^+$ | | | | | | | | <10 (32) | | |
| $PhHg^+$ | | | | | | | | <10 (32) | | |
| $CH_3HgCl$ | 80 (28) | 30 (28) | <10 (28) | 5650 (29) | 180–200 (30,33) | 60–110 (33) | <10–11,500 (28,29,32, 34,8,18, 36–41) 8300 (34) | 10–910 (28–32) | 110–880 (31,33) | 60–1460 (30,33) |

| Compound | Rainbow Trout | Whale | Shark | Swordfish | Octopus | Squid | Rockfish |
|---|---|---|---|---|---|---|---|
| $CH_3HgCl$ | 50–1970 (30) | 560–1090 (30) | <10 (32) 5410 (35) | 570–1010 (28,36) 400–3170 (28,29) | <10 (28) | <10 (28) | 100–1000 (33) |

**Table 2.16.** Concentrations of organometallic compounds and corresponding inorganic metal ions found in fish ($\mu g\ kg^{-1}$).

|  | Organometal | | Inorganic Metal | |
|---|---|---|---|---|
| Element | Minimum | Maximum | Minimum | Maximum |
| Lead | 1500 (m) | 1600 (a) | 120 (i) | 1360 (i) |
|  | 1 (b) | 5500 (b) |  |  |
|  | 5100 (m) | 18,900 (m) |  |  |
|  | 7700 (c) | 12,600 (c) |  |  |
| Mercury | 100 (d) | 900 (d) | 90 (j) | 2400 (m) |
|  | 600 (e) | 1500 (e) |  |  |
|  | 50 (f) | 2000 (f) |  |  |
|  | 600 (g) | 1100 (g) |  |  |
|  |  | 8400 (h) |  |  |
| Arsenic | — | — | 1100 (k) | 2900 (k) |
| Copper | — | — | 390 (l) | 3500 (l) |
|  |  |  | 500 (o) | 220 (o) |
| Nickel | — | — | 150 (i) | 200 (i) |
| Selenium | — | — | 190 (n) | 550 (n) |
| Manganese | — | — | 220 (i) | 1630 (i) |

(a) pike, (b) bass, (c) white sucker, (d) whiting, (e) sardine, (f) turbot, (g) halibut, (h) Salmon, (i) mullet, (j) chub, (k) herring, (l) flathead, (m) carp, (n) shark, (o) rainbow trout.

in Table 2.17, which shows the concentrations of metals in the organ can be up to 30 times higher than in the fish tissue.

Richinon et al.[42] have discussed the factors that might govern the uptake of mercury by fish in acid stressed lakes. It was concluded that mercury cycling and uptake in aquatic systems were governed by a variety of interconnecting and sometimes covarying factors, the selective importance of which could differ from lake to lake.

**Table 2.17.** Concentrations of metallic ions in fish tissues and organs.

| Element | Fish tissue analysis ($\mu g\ kg^{-1}$) | | Fish organ analysis ($\mu g\ kg^{-1}$) | | Concentration factors, Maximum Value in Organ ($\mu g\ kg^{-1}$) |
|---|---|---|---|---|---|
|  | Minimum | Maximum | Minimum | Maximum | Maximum Value in Tissue ($\mu g\ kg^{-1}$) |
| Lead | 120 (a) | 1360 (b) | 120 (A) | 36000 (B) | 26.4 |
| Mercury | 90 (c) | 2400 (d) | — | — | — |
| Arsenic | 1100 (e) | 2900 (f) | — | — | — |
| Copper | 390 (g) | 3500 (h) | 600 (D) | 48 000 (l,C) | 13.7 |
|  | 530 (i) | 2180 (j) |  | 62 000 (m,C) | 28.4 |
| Nickel | 150 (i) | 200 (i) | 340 (n,B) | 1900 (n,B) | 9.5 |
| Selenium | 190 (k) | 550 (k) | — | — | — |
| Manganese | 220 (a) | 1630 (j) | — | — | — |

(a) striped mullet, (b) grey mullet, (c) Chub, crappie, (d) carp, (e) herring, (f) tuna, (g) flathead, (h) cray fish (i) rainbow trout, (j) sardine, (k) shark, (l) perch, (m) white fish, (n) trout. (A) muscle, (B) kidney, (C) liver, (D) gill.

## 2.4.2 CRUSTACEA

Available information on the concentrations of organometallic compounds and the corresponding inorganic metal ions is given in Table 2.18. Again, as in the case of fish, determined values vary over a wide range and certainly include in this range concentrations at which adverse effects or mortalities could occur and when the suitability of the creature as an item of human diet would be queried. In particular, the concentrations of organolead compound and inorganic lead found in crustacea are lower than those found in fish (Table 2.19)

Data has been presented on the concentrations of ionic alkyl lead compounds in saltmarsh periwinkles (*Littorina irrorata*) collected in Maryland and Virginia[43]. Male periwinkles accumulated higher concentrations of several alkyl lead species than females.

Bailey and Davies[44] determined tributyl tin concentrations in dogwhelk samples taken at various locations in Sullom Voe, Shetland. These ranged from values of 0.1 mg kg$^{-1}$ inside Sullom Voe down to less than 0.03 mg kg$^{-1}$ in Yell Sound. Concentrations of 0.02 to 0.03 mg kg$^{-1}$ were found in edible tissue of queen scollops inside the Voe but tin was rarely detected in commercial shellfish outside the Voe. Only very low concentrations of tributyl tin ($\sim$ 2 ng L$^{-1}$) were found in a small proportion of sea water samples taken in the area.

**Table 2.18.** Occurrence of organometallic compounds and corresponding inorganic metal ions in crustacea.

|  | Organometallic compounds (µg kg$^{-1}$) | | Inorganic compounds (µg kg$^{-1}$) | |
| --- | --- | --- | --- | --- |
|  | Minimum | Maximum | Minimum | Maximum |
| Lead | <10 | 50 | 480 (A) | 2800 (D) |
| Mercury | — | — | 20 (A) | 310 (E) |
| Arsenic | 1400 (G) | 23 000 (H) | 20 (B) | 180 (F) |
| Copper | — | — | 750 (A) | 2650 (A) |
| Tin | 30 (F) | 100 (F) | — | — |
| Selenium | 20 (C) | 30 (C) | 710 (C) | 6700 (E) |

(A) muscle, (B) king prawn, (C) scallop, (D) crab, (E) lobster, (F) whelk, (G) prawn, (H) mollusc.

**Table 2.19.** Comparison of ranges of concentrations of organolead compounds found in crustacea and fish.

| Compound | In crustacea (µg kg$^{-1}$)[49] | In fish (µg kg$^{-1}$) (Table 2.14) |
| --- | --- | --- |
| R$_4$Pb | <20 | 1117–9360* |
| R$_3$Pb$^+$ | <10–50 | 63–2640* |
| R$_2$Pb$^{2+}$ | <10 | 450–4660* |
| Pb$^{2+}$ | 1100–1800 | 250–4300 |

*ethyl plus methyl compounds

Minchin et al.[45] studied changes in shellfish populations, including scallops (*Pecten maximus*) and flame shells (*Lima hians*) in Mulroy Bay, Eire, before, during and after the extensive use of organotin compounds in net-dips on salmonid farms in the area. The results were compared with data from other areas on the west and south coasts of Eire. Tabulated data are included on estimated numbers of scallops in each area, concentrations of tributyltin in molluscan tissues, and shell distortions and tributyltin concentrations in oysters, Although use of tributyltin was discontinued in Spring 1985, relatively high concentrations were found in adult scallop tissues, but settlements of scallops had reappeared. The concentrations in mussels and flame shells were relatively low, and some mussel settlement had occurred in Mulroy Bay, but flame shells had failed to settle since 1983, and the remaining population was endangered. Since April 1987, the use of organotin antifouling compounds on boats and other aquatic structures had been illegal in Eire.

## 2.5 THE PHYTOPLANKTON AND WEED ECOSYSTEM

High concentrations of inorganic metals and some organometallic compounds are found in algae and phytoplankton in rivers where pollution is occurring. This is important in the sense that these are the foodstuffs of fish and crustacea who by eating them increase their body burden of these contaminants. In Table 2.20 are compared typical values of inorganic copper and arsenic and organolead

**Table 2.20.** Comparison of inorganic metal and organometallic contents ($\mu g\ kg^{-1}$) of algae, phytoplankton, fish and crustacea (references in parentheses).

| Inorganic metals | Organometallic concentration in | | | | | |
|---|---|---|---|---|---|---|
| | Compounds | Algae | Marine crown algae (laminari) | Phytoplankton up to $\mu g\ kg^{-1}$ | Fish | Crustacea |
| Copper | — | 660 000 | | 40 000 | 3500 | 2600 |
| Arsenic | — | 56 000 | | 2900 | 180 | |
| | | 200–600[47](a) | | | | |
| | | 7600– 15 600[47](b) | 40 300– 89 700[47] | — | — | — |
| | Organolead Compounds | | | | | |
| | Me$_3$EtPb | — | | 38 | <1 (27) | — |
| | Me$_2$Et$_2$Pb | — | | 1500 | 57–1430 (27) | — |
| | MeEt$_3$Pb | — | | 3610 | 140–290 (27) | — |
| | Et$_4$Pb | — | | 16 500 | 780–7500 (27) | — |
| | Et$_3$Pb$^+$ | — | | 560 | 53–3430 (27) | — |
| | Et$_2$Pb$^{2+}$ | — | | 110 | 90–4300 (27) | — |
| | Total | — | | 22 300 | 1400–21 100 (27) | — |

[a] As monomethyl arsenic acid.
[b] As dimethyl arsenic acid.

compounds in algae, phytoplankton, fish and crustacea. These results confirm that, in general contaminant levels are higher in algae and phytoplankton than they are in fish and crustacea.

Some concentrations of inorganic elements in algae and freshwater phytoplankton are given in Table 2.21. These range from 20 000 µg kg$^{-1}$ of arsenic found in algae to as high as 4 170 000 µg kg$^{-1}$ of manganese found in algae.

**Table 2.21.** Comparison of inorganic metal contents of river waters, phytoplankton and algae.

| Element | Concentration in | | | Mean concentration in algae $\frac{B}{A}$ | Mean concentration in phytoplankton $\frac{C}{A}$ |
|---|---|---|---|---|---|
| | River water $A(\mu g\ L^{-1})$ | Algae $B(\mu g\ kg^{-1})$ | Freshwater phytoplankton $C(\mu g\ kg^{-1})$ | | |
| Mercury | 0.009–13.0 | — | 31 200–81 000 | — | 8615 |
| Arsenic | 0.42–490 | 20 000–56 100 | — | 155 | — |
| Copper | 0.11–200 | 50 000–660 000 | 40 000 | 3550 | 400 |
| Manganese | — | 230 000–4 170 000 | 230 000–250 000 | — | — |

Algae and phytoplankton have the property of concentrating in their tissues elements present in the water in which they grow. This has two advantages as far as the environmental chemist is concerned. By analysing the algae or phytoplankton instead of the water in which they grow much more sensitivity may be achieved in the analysis. Secondly, whereas analysis of the water provides only a spot check on its composition at a particular point in time, the metal content of the algae or phytoplankton is the result of exposure to the metal over a prolonged period of time and hence provides an integrated value for the level of metal present in the water over a length of time.

Some idea of the concentrations involved may be gained by calculating the concentration factor

$$\frac{\text{mean concentration in algae or phytoplankton }(\mu g\ kg^{-1})}{\text{mean concentration in water }(\mu g\ L^{-1})}$$

These range from 155 in the case of inorganic arsenic in algae (i.e. the algae contains approximately 155 times the concentration of arsenic than that present in the water in which it grows), to as high as 8615 in the case of mercury in freshwater phytoplankton. Thus, in the case of mercury the detection of 5 µg kg$^{-1}$ mercury in the phytoplankton implies that the mean mercury content of the water in which it exists is 0.000 58 (i.e. 5/8615) µg L$^{-1}$, i.e. 0.58 ng L$^{-1}$, an analytical detection level which it would be either extremely difficult or impossible to achieve by direct analysis of the water.

These concentration effects occur in the case of organometallic compounds as well as inorganic metals. Thus, in the case of organolead compounds, similar levels of organolead have been found in phytoplankton to those found in fish tissues (Table 2.22). In both cases the organolead levels are considerably higher than those found in the water which the fish or phytoplankton live.

Table 2.22. Comparison of concentrations of organolead compounds found in phytoplankton and fish. From Capelli et al.[34] Reproduced by permission of Royal Society of Chemistry.

| Compound | In fish ($\mu g\ kg^{-1}$) (from Table 2.20) | In phytoplankton[34] ($\mu g\ kg^{-1}$) |
|---|---|---|
| $Me_3EtPb$ | <1 | 38 |
| $Me_2Et_2Pb$ | 57–1430 | 1500 |
| $MeEt_3Pb$ | 140–290 | 3610 |
| $Et_4Pb$ | 780–7500 | 16 500 |
| $EtPb^+$ | 53–3430 | 560 |
| $EtPb^{2+}$ | 90–4300 | 110 |
| Total | 1400–21 100 | 22 300 |

## 2.6 THE SOIL AND CROPS ECOSYSTEM

A proportion of the inorganic metals (some of which are potentially capable of forming organometallic compounds by biosynthetic processes) that are present in soil enter grass and vegetable and grain crops on the land. This represents a potential hazard to farm animals and to man who ingest the animals and crops.

Whilst some of these metals are naturally occurring, i.e. may have a geological origin, the remainder are present due to man-made pollution such as airborne pollutants and the use of agrichemicals, sewage sludge added to the land as a fertiliser, or, possibly, artificial fertilisers. In Table 2.23 are shown concentration

Table 2.23. Relationship between inorganic metal ion concentrations in land and crops.

| Metal ion | Crops ($\mu g\ kg^{-1}$) | Soil ($\mu g\ kg^{-1}$) |
|---|---|---|
| Lead | 30–3200 (1600) | 30–710 000 |
| Mercury | 150 | 30–3300 |
| Tin | — | 1500–7100 |
| Arsenic | 8–1000 | 6000–1 375 000 |
| Antimony | <2–40 | 600–66 000 |
| Selenium | 2–1900 | 10–111 000 |
| Copper | 100–13 700 (1600) | 1000–240 000 |
| Nickel | 10–7500 (1600) | 3000–74 000 |
| Manganese | 500–173 000 | 188 000–2 750 000 |

Acceptable values in crops in parentheses

ranges of inorganic metal ions (some at least of which are capable of forming organometallic compounds by biosynthetic processes) that can occur in land and crops.

As would be expected the concentrations of such inorganic ions in crops are lower than in soil. At their maximum observed level in crops the concentrations of lead, copper and nickel exceed recommended acceptable values of 1600 µg kg$^{-1}$ by factors of approximately two (lead) to eight (copper). It is essential, therefore, to be aware of toxicant levels in soils and crops if unforeseen consequences for the health of man and animals are to be avoided.

## 2.6.1 CONSEQUENCES OF APPLYING SEWAGE SLUDGE TO LAND

We are only on the periphery of understanding what biotransformations from inorganic metals to organometallic compounds can occur when metal-contaminated sewage sludge is disposed of as a fertiliser on agricultural land or in the oceans.

Some 45% of the 1.23 million tonnes dry sewage sludge produced annually in the UK and Wales is used as an agricultural fertiliser, the remainder being directly disposed of in rivers or as outfalls to the sea. This sludge, if it originates in a sewage works that also handles industrial effluents containing metals and does not treat such effluents separately, can contain high levels of metal. The disposal of such metal-contaminated sludge to land can be harmful to plant and/or animal life and, consequently, to man who consumes animals and crops. This is illustrated in the case of lead in Table 2.24. As mentioned above, the presence of inorganic metal ions in sewage sludge creates a potential for the formation of organometallic compounds by biosynthetic processes.

Typical concentrations of inorganic metal ions found in typical agricultural soils and neat sewage sludge are compared in Table 2.25. At their maximum observed concentrations appreciably higher levels of all these metals are present in the sludges than in the soil. Hence, there is a possibility that sewage sludge treated soils will contain more than the acceptable levels of metals for crop growing.

At the lower end of their concentration ranges quoted in Table 2.25, the lead, copper and manganese content of the sewage treated soil is acceptable for crop growing whilst at the higher end of their concentration range they are not acceptable.

**Table 2.24.** Metal contents (µg kg$^{-1}$) sewage sludge treated land compared with metal contents of untreated good quality farm land.

| Element | Untreated scottish arable soil | Untreated canadian soil | Sewage treated soil | Acceptable value for crop growing |
|---|---|---|---|---|
| Lead | 8000–44 000 | 4290–9800 | — | 100 000 |
| Copper | 3100–26 000 | 4300–56 000 | 1000–1 170 000 | 100 000 |
| Nickel | 3800–26 000 | <2000–74 000 | — | 100 000 |

**Table 2.25.** Metal content ($\mu g\ kg^{-1}$) of soil and sewage sludge.

| Element | Sewage treated soil | Neat sewage sludge |
|---|---|---|
| Lead | 30–710 000 (100 000) | 10 000–3 500 000 |
| Mercury | 30–3300 | 1400–26 000 |
| Tin | 1500–7100 | 3500–66 000 |
| Arsenic | 6000–1 375 000 | 850–150 000 |
| Antimony | 600–66 000 | 2800–605 000 |
| Selenium | 10–111 000 | 140–10 000 |
| Copper | 1000–240 000 (100 000) | 74 000–3 650 000 |
| Nickel | 3000–74 000 (100 000) | 26 000–940 000 |
| Manganese | 188 000–2 750 000 (400 000) | 120 000–1 540 000 |

Acceptable values for crop growing in parentheses

To meet the recommended maximum advisable concentrations for metals including organometallic compounds in soil (viz. 100 000 $\mu g\ kg^{-1}$ for lead, copper and nickel and 400 000 $\mu g\ kg^{-1}$ for manganese) controls must be exerted on the composition of sewage sludge and on its application rate to land. Thus sludge is usually applied to land at the rate of 100–500 tonne per hectare. One hectare of land 20 cm deep weighs approximately 3000 tonne. Therefore, the application of 100 tonne dry sludge to the top 20 cm of a hectare of land represents a 3.3% addition by weight. If the sludge contained $10^6$ $\mu g\ kg^{-1}$ of lead then the metal content of the soil will increase by 3.3%, of $10^6$ i.e. 33 000 $\mu g\ kg^{-1}$. If the land before application of the sludge contained 70 000 $\mu g\ kg^{-1}$ lead, then after application this would increase to 103 000 $\mu g\ kg^{-1}$, i.e. above the maximum acceptable concentration of 100 000 $\mu g\ kg^{-1}$ lead. Subsequent applications of sewage sludge would take the lead content of the soil well above the acceptable limit. Applications of sewage to land is therefore to be viewed with extreme caution.

## 2.7 POTABLE WATER

Concentrations found in a wide range of potable waters of inorganic metals of possible environmental concern from the point of view of either occurring as organometallic compounds or being converted to organometallic compounds by biotransformation are listed in Table 2.26. Of these elements only mercury has been encountered in its organometallic form (alkylmercury) in potable water. The concentrations of alkymercury compounds found (0.0007–0.003 $\mu g\ L^{-1}$) are well below the limit of any environmental concern. Thus the US Safe Drinking Water Act[5,24] allows a maximum of 0.5 $\mu g\ L^{-1}$ total mercury including alkylmercury in potable water. The World Health Organisation has suggested a maximum daily intake by humans via potable water of 43 $\mu g$ total inorganic plus organometallic mercury and 29 $\mu g$ alkylmercury. Thus, even if the water contained the WHO limit of 0.5 $\mu g\ L^{-1}$ total mercury the average daily water consumption would have to be 86 L before there is cause for concern.

Table 2.26. Concentration ranges of inorganic metal ions found in potable water.

| Element | Concentration range ($\mu g\ L^{-1}$) | WHO Limit ($\mu g\ L^{-1}$) | US Safe Drinking Water Act[37,48] |
|---|---|---|---|
| Mercury | <0.001–0.010 | 0.5 | 2 |
| Lead | <0.6–565 | — | — |
| Selenium | <10 | — | — |
| Manganese | 1.9–20.5 | — | — |
| Arsenic | 5.4–11.4 | — | — |
| Antimony | <13 | — | — |
| Copper | <3–945 | 50 | — |
| Nickel | 0.19–6.1 | — | — |

At the actual maximum concentration of alkylmercury of 0.003 $\mu g\ L^{-1}$ occurring in potable water a daily consumption of 10 000 L would be necessary before the WHO limit of 0.5 $\mu g\ L^{-1}$ were exceeded.

The daily intake of mercury by humans from potable water is considerably less than the total daily intake of 1–20 $\mu g\ L^{-1}$ from foods.

Exceedingly low concentrations of alkyl tin compounds occur in potable water. No other types of organometallic compounds have been detected.

## 2.8 THE ATMOSPHERIC ECOSYSTEM

Due mainly to emissions from metal industry smoke stacks and automobile exhausts the concentration of metals in the atmosphere, and hence in rain and other forms of precipitation, is not inappreciable.

The concentrations of organometallic compounds and the corresponding inorganic metals that have, to date, been found in the atmosphere are reviewed in Table 2.27. Some 24% (i.e. 16 956 tonne) of the total annual metal pollution load of the North Sea (i.e. 706 500 tonne) arrives in the form of atmospheric pollution.

Table 2.27. Concentrations of organometallic compounds and inorganic metal ions found in the atmosphere (references in parentheses).

| Element | Organometallic compounds ($\mu g\ L^{-1}$) | Inorganic metals ($\mu g\ L^{-1}$) |
|---|---|---|
| Lead | — | 0.05–40 (Snow/rain) |
| Mercury | 0.009 (5) (Rain) | 0.014–0.025 (Rain) |
| Tin | <0.001–0.006 (2,3) (Rain) | 0.025 (Rain) |
| Antimony | — | 0.002–0.089 (Rain) |
| Selenium | — | 0.005–0.025 (Snow) |
| Copper | — | 0.02–0.097 (Snow) |
| Nickel | — | 5.0 (Rain) |

It has been estimated that approximately $10^{12}$ tonnes (i.e. $10^{15}$ L) of rain falls annually on the North Sea. If the inorganic plus organic lead content of this rain is taken to be in the range 0.05–40 μg L$^{-1}$ (Table 2.27), this would represent an annual concentration of some 50–40 000 tonne of total lead of which, say, 10% i.e. 5–4000 tonne is organolead. As well as contaminating the oceans such atmospheric contributions will also contaminate land and rivers.

## REFERENCES

1. Braman R.S., Tompkins M.A. *Analytical Chemistry* **51** 12 (1979)
2. Landy M.P. *Anal. Chim. Acta* **121** 39 (1980)
3. Muller M.D. *Analytical Chemistry* **59** 617 (1987)
4. Kiemencz A.M., Kloosterboer J.G. *Analytical Chemistry* **48** 575 (1976)
5. Minagawa K., Takizawa A., Kifune I. *Anal. Chim. Acta* **115** 103 (1980)
6. Potter H.R., Jarview A.W., Markell R.N. *Water Pollution Control* **76** 123 (1977)
7. Andren A.W., Harris R.C. *Nature (London)* **245** 256 (1973)
8. Matsunaga K., Takahaishi S. *Anal. Chim. Acta* **87** 487 (1976)
9. Gilmore C.C., Tutte J.H., Meons J.C. In *Marine and Estuary Geochemistry* Siglio A.C., Haltori A. (Editors), Lewis Publishers, Chelsea, pp239–258 (1985)
10. Gilmour C.C., Tuttle J.H. *Analytical Chemistry* **58** 1848 (1986)
11. Shopintsev B.A. *Oceanography* **6** 361 (1966)
12. Williams P.J. 'Determination of organic components' In *Chemical Oceanography* Riley J.P., Skirrow G. (Editors) Volume 3. Academic Press, New York, pp443–477 (1975)
13. Skopintsev B.A. *Oceanography* **16** 630 (1976)
14. Wangersky P.J., Zika R.G. *The Analysis of Organic Compounds in Seawater Report 3*, NRCC 16566 Marine Analytical Chemistry Standards Programme (1978)
15. Van Hall C.E., Barth D., Stenger V.A. *Water Sewage* **111** 266 (1964)
16. Van Hall C.E., Barth D., Stenger V.A. *Analytical Chemistry* **37** 769 (1965)
17. Van Hall C.E., Stenger V.A. *Analytical Chemistry* **39** 502 (1967)
18. Langsten W.J., Burt G.R., Mingjiang Z. *Marine Pollution Bulletin* **18** 634 (1987)
19. Davies I.M. Graham W.C., Pirie S.M. *Marine Chemistry* **7** 11 (1979)
20. Haywood M.G., Riley J.P. *Analytical Chemistry* **85** 219 (1976)
21. Ram S.N., Sathyanesan A.G. *Environmental Bulletin* **47** 135 (1987)
22. Jackson J.A., Blair W.R., Brinckman F.E., Iveson W.P. *Environmental Science and Technology* **16** 111 (1982)
23. Valkirs A.O., Seligman P.F., Strong P.M. *Marine Pollution Bulletin* **17** 319 (1986)
24. Ebdon L., Alonso G. *Analyst (London)* **112** 1551 (1987)
25. Jirka A.M., Carter M.J. *Analytical Chemistry* **50** 91 (1978)
26. Hodge W.F., Seidel S.L., Goldberg D. *Analytical Chemistry* **51** 1256 (1979)
27. Chau J.K., Wong P.T.S., Bengent G.A., Dunn J.L. *Anal. Chem.* **56** 271 (1984)
28. Holak W. *Analyst (London)* **107** 1457 (1982)
29. Shum G.T.C., Freeman H.C., Uthe J.F. *Analytical Chemistry* **51** 414 (1971)
30. Uthe J. *Fisheries Research Board*, Canada
31. Farrington J.W., Teal J.M., Quinn J.G., Wade T., Burns K.A. *Bulletin of Environmental Contamination and Toxicology* **10** 129 (1973)
32. MacCrehan W.A., Durst R.A. *Analytical Chemistry* **50** 2108 (1978)
33. Matsunaga K., Takahasi S. *Anal. Chim. Acta* **87** 487 (1976)
34. Capelli R., Fezia C., Franchi A. *Analyst (London)* **104** 1197 (1979)
35. Dietz E.A., Singley K.F. *Analytical Chemistry* **51** 1809 (1979)

36. Fielding M., McLaughlin K., Steel C. *Water Research Centre Enquiry Report ER532*, August, 1977, Water Research Centre, Stevenage Laboratory, Elder Way, Stevenage, Herts (1977)
37. Westoo F. *Anal. Chem. Scan.* **21** 1790 (1967)
38. Laughlin R.B., Linden O. *Ambio* **16** 252 (1987)
39. Callum G.I., Ferguson M.M., Lenchan J.M.A. *Analyst (London)* **106** 1009 (1981)
40. Analytical Methods Committee Society for Analytical Chemistry, London, *Analyst (London)* **102** 769 (1977)
41. Uthe J.F., Solomon J. Griff B. *J. Association of Official Analytical Chemists* **55** 583 (1972)
42. Richman L.A., Wren C.D., Stokes P.M. *Water, Air and Soil Pollution* **37** 465 (1988)
43. Krishnan K., Marshall W.D., Hatch W.I. *Environmental Science and Technology* **22** 806 (1988)
44. Bailey S.K., Davies I.M. *Environmental Pollution* **55** 161 (1988)
45. Minchin D., Duggan C.B., King W. *Marine Pollution Bulletin* **18** 604 (1987)
46. White J.N.C., Englar J.R. *Botanica Marina* **26** 159 (1983)
47. Maher W.A. *Anal. Chim. Acta* **126** 157 (1981)
48. Jones P., Nickless G. *Analyst (London)* **103** 1121 (1978)
49. Stratton G.W. *Bulletin of Environmental Contamination and Toxicology* **38** 1012 (1987)
50. Maguire R.J., Tkracz R.J. *Water Pollution Research Journal of Canada* **22** 227 (1987)

# CHAPTER 3
# Toxic Effects of Organometallic Compounds

## 3.1 TOXICITY DATA

### 3.1.1 INTRODUCTION

A fair amount of information is now available on the risk of adverse effects on health or mortality to various water creatures exposed to known concentrations of inorganic metals for stated periods of time. Some data is quoted in Table 3.1 showing the creatures for which mortalities, or at least, adverse effects would be expected as the concentrations of these metals approach the higher limits that have been detected in the aqueous environment. Unfortunately, similar detailed information for organometallic compounds is sparce. What is known is discussed in this chapter.

The four types of organometallic compounds that occur most frequently in the environment and for which some toxicity data is available are those of mercury, lead, tin and arsenic. These can originate in the ecosystem either as man-made pollutants or by microorganism (molds, bacteria) induced biomethylation of inorganic metals in sediments, fish, marine invertebrates or soil. Little is known of the pathways by which these processes occur. Little or no toxicity data is available for the less commonly encountered organic compounds of other elements such as selenium, copper etc.

### 3.1.2 ORGANOMERCURY COMPOUNDS

Certain organomercury compounds, such as $HgMe_2$ and $HgMeCl$, are more toxic than elemental mercury (Uthe and Armstrong[1]), and inorganic mercury forms and, when present in the environment, may cause serious illness in extremely polluted areas (Backmann[2]).

The toxicity of mercury is related to its chemical form. Liquid mercury appears to have no effect but mercury vapour is readily adsorbed by humans causing brain damage. Mercurous salts are relatively non toxic compared to mercuric salts because of their lower solubility in body fluids.

A growing public interest in environmental quality has lead to the development of analytical techniques for the monitoring of environmental pollutants. Due to its chronic toxicity and its tendency to bioaccumulate, mercury is of prime interest.

**Table 3.1.** Toxicity data on inorganic mercury, lead, nickel and copper (4–14 days exposure). (A) Concentration above which mortalities can occur, concentrations found in (B) River water, (C) Coastal/estuary water, (D) Seawater, all in µg L$^{-1}$.

| Element Creature | Mercury A | B | C | D | Lead A | B | C | D | Copper A | B | C | D | Nickel A | B | C | D |
|---|---|---|---|---|---|---|---|---|---|---|---|---|---|---|---|---|
| Annelids (adult) | 100 | 0.009–13.0 | 0.00002–15.1 | 0.002–0.078 | 100 | 0.13–60 | 0.035–7.44 | 0.00004–9.0 | 10 | 0.11–200* | 0.069–20.0 | 0.0063–6.8 | 10.000 | 1.5–4.5 | 0.2–5.33 | 0.15–0.93 |
| Annelids (larval) | — | — | — | — | — | — | — | — | 10 | "** | " | " | — | — | — | — |
| Bivalve molluscs (adult) | 0.1 | "** | "** | " | 1000 | " | " | " | 1 | "** | " | " | 10.000 | " | " | " |
| Bivalve molluscs (larval) | 1 | "** | "** | " | 100 | " | " | " | 1 | "** | " | " | 100 | " | " | " |
| Crustacea (adult) | 10 | "** | "** | " | 1000 | " | " | " | 10 | "** | " | " | 100 | " | " | " |
| Crustacea (larval) | 1 | "** | "** | " | 100 | " | " | " | 10 | "** | " | " | 1000 | " | " | " |
| Echinoderm (adult) | — | — | — | — | 1000 | " | " | " | 10 | "** | " | " | 10 | " | " | " |
| Gastropods | — | — | — | — | 100 | " | " | — | 10 | "** | " | " | 10.000 | " | " | " |
| Hydrozoans | — | — | — | — | 1000 | " | " | — | 10 | "** | " | " | — | — | — | — |
| Fish (adult) | 10 | "** | "** | — | 1000 | " | " | — | 100 | "** | " | " | 100 | " | " | — |
| Fish (larval) | — | — | — | — | — | — | — | — | 10 | "** | " | " | — | — | — | — |

*Mortalities can occur.

Being volatile in its organic and elemental forms, mercury is well dispersed in the atmosphere and the aqueous and terrestrial ecosystems.

The activity of certain bacteria, molds and enzymes in the soil or sediment can produce methylated mercury from elemental or inorganic mercury (Ridler et al.[3], Imura et al.[4], Wood et al.[5], Jenson and Jernelov[6]). The organic mercury compounds produced, primarily dimethylmercury and methylmercury halides, are potentially more toxic than inorganic mercury forms. Therefore, recent studies of environmental mercury have been concerned with its chemical speciation to determine not only the amounts of mercury present but the chemical forms as well. More extensive data in this area will assist in determining the role of organic mercury in the global cycling of the element.

The interest in mercury contamination, and particularly in the organic mercury compounds, is a direct reflection of the toxicity of these compounds to man. Some idea of the proliferation of work can be derived from the reviews of Krenkel[7], Robinson and Scott[8] and Uthe and Armstrong[9].

All forms of mercury are potentially harmful to biota, but monomethyl and dimethyl mercury are particularly neurotoxic. The liphophilic nature of the latter compounds allow them to be concentrated in higher trophic levels and the effects of this biomagnification can be catastrophic[10].

Environmental factors influence the net amount of methyl mercury in an ecosystem by shifting the equilibrium of the opposing methylation and demethylation reactions. This may be represented as follows:-

$$Hg \rightleftharpoons MeHg^+ \rightleftharpoons Me_2Hg \quad \substack{\longleftarrow \text{ demethylation (non-specific hydrolytic} \\ \text{ and reductive enzyme process)} \\ \longrightarrow \text{ Methylation } (Hg^{2+} \text{ interference with} \\ \text{ biochemical C-1 transfer)}}$$

Methylation is the result of mercuric ion ($Hg^{2+}$) interference with biochemical C-1 transfer reactions[11]. Demethylation is brought about by non-specific hydrolytic and reductive enzyme processes[12-14]. The biotic and abiotic influences that govern the rates at which these processes occur are not completely understood.

Although much of the early work on the cycling of mercury pollutants has been performed in freshwater environments, estuaries are also subject to anthropogenic mercury pollution[15]. A strong negative correlation exists between the salinity of anaerobic sediments and their ability to form methymercury from $Hg^{2+}$. As an explanation for this negative correlation the theory was advanced that sulphide, derived by microbial reduction of sea salt sulphate, interferes with $Hg^{2+}$ methylation by forming mercuric sulphide which is not readily methylated[16-19]. There are several reports in the literature concerning the methylation of $Hg^{2+}$ by methylcobalamin[3,12,20].

The synthesis of methylmercury compounds form inorganic mercury by microorganisms, molds and enzymes has been investigated by several workers[3,4,5,6,20,21-26]. This biological methylation of mercury compounds provides an explanation for the fact that $CH_3Hg^+$ is found in fish, even if all known sources of mercury in the environment are in the form of inorganic mercury or phenyl mercury. The formation of the volatile $CH_3HgCH_3$ (b.p. 94°C) may be a factor in the redistribution of mercury from aqueous industrial wastes. The process of methylation is fundamental to a knowledge of the turnover of mercury; it may be significant in the uptake and distribution of mercury in fish and in the mobilisation of mercury from deposits in bottom sediments into the general environment.

It has been reported that organomercury compounds are significantly concentrated in fish predominantly as methylmercury compounds[28-36].

The synthesis of methylmercury compounds by microorganisms in freshwater sediments have been investigated by some workers (Jensen and Jernelov[6], Wood et al.[5]).

Although methylmercury has been found in aquatic organisms, its origin is not in all cases clearly known. It is generally assumed that methylmercury exists in natural waters and that the organisms concentrate it, because it has been detected in many aquatic organisms.

Marine fish embryos and also marine crustacea and crab undergo adverse effects (damage and poor hatching) when exposed to 30–70 µg $L^{-1}$ inorganic mercury in sea water for 4 to 30 days. Similar effects occur in the case of marine molluscs.

Severe adverse effects (weight reduction and poor spawning) occur when non-marine fish are exposed to as little as 1 µg $L^{-1}$ methylmercury in water for 30 days. 3 µg $L^{-1}$ mercury, as methylmercury chloride, in seawater or rivers caused 100% mortality in fish. The toxic effects of organic and inorganic mercury salts to fish are similar in river water and seawater.

Ram and Sathyanesan[37] have studied the histopathological and biochemical changes in the liver of teleost fish *Channa punctates* induced by a mercurial fungicide.

Twenty individuals of *C. punctatus* from two age groups (adults and young of 6 months old) were divided into 2 batches of 10 specimens. Batch 1 of both age groups was subjected to a toxicologically safe dose (0.20 mg $^{-1}$ of the commercial formulation of organomercury fungicide, Emisan (methoxy ethyl mercuric chloride). Batch 2 of each age group was kept as an untreated control. After 6 months exposure, the fish were killed and the livers examined histochemically. In both the treated groups, liver histology showed various abnormalities, including hyperplasia, nuclear pycnosis, fatty necrosis, and degeneration of hepatocytes leading to a tumour and syncytium formation, which are indicative of carcinogenesis. Young fish also showed blood vessel congestion and oedema. Biochemical changes (reduction in hepatosomatic index, total protein and lipid, elevation in cholesterol and acid and alkaline phosphatase) occurred in both exposed groups but were more pronounced in young than in adult fish. Emisan, even in small concentrations, was capable of inducing severe physio-metabolic dysfunction leading to death in *Channa punctatus*.

Kirubagaran and Joy[38] examined the toxic effects of three organomercury compounds on the survival and histology of the kidneys of the catfish *Clarius batrachus L*. In acute toxicity tests with *Clarias batrachus* (24,48,72,96 h $LC_{50}$s calculated according to Spearman–Karber with trimming at 0 and 10%) methylmercuric chloride, mercuric chloride and Emisan 6 (methoxyethyl mercury chloride) had mean 96 h $LC_{50}$s (0% trimming) of 430, 507 and 432 µg $L^{-1}$ respectively. In subsequent experiments with *C. batrachus*, concentrations of methylmercuric chloride, mercuric chloride and Emisan 6, respectively, were 200, 125 and 1500 µg $L^{-1}$ (14, 28 days exposure), or 40, 50 and 500 µg $L^{-1}$ (90, 180 days exposure). With the exception of the 180 day treatment groups, kidney histology of mercury exposed fish revealed increases in the proximal tubule diameter and secretory material.

Czuba *et al.*[39] exposed cell suspension cultures of *D. carota* to methylmercury (0–6 µg per mL) for 1,3 or 24 h. Mircotubule arrays were unaffected by 1000 µg $L^{-1}$ (all exposures) and severely disrupted by 6000 µg $L^{-1}$ (all exposures). The degree of disruption caused by other treatments was dependent on exposure duration and concentration. Experimental evidence indicated that microtubule disruption might be a secondary effect resulting from methylmercury induced perturbations in protein and carbohydrate metabolism and disruptions in redox energy related processes.

The concentration of organomercury present in sea and coastal water (up to 0.06 µg $L^{-1}$ [40], Table 2.5) probably presents little risk to the health of most types of water creatures, although this cannot be taken as a general rule. Thus, 0.2 µg $L^{-1}$ organomercury causes severe mortalities in crayfish during three days exposure. The higher levels in the range 0.006–1.15 µg $L^{-1}$ of organomercury in rivers[41–42] (Table 2.1) will certainly be harmful to fish and many types of water creatures. These higher levels of organomercury, say above 0.05 µg $L^{-1}$, are more likely to occur in localised areas, i.e. near highly contaminated industrial outfalls to rivers, estuaries and coastal waters, e.g. the Minimata Bay, Japan, incident.

Regarding concentrations of organomercury in water creature tissues, these have been observed to range from <10 µg $kg^{-1}$ (white tuna, shark, octopus, squid, turbot) to as high as 8300 µg $kg^{-1}$ in red tuna[43].

The concentration of organomercury that is harmful to water creatures is not necessarily the same as the concentration that is harmful to man who eats the fish, i.e. a fish that, although it has traces of organomercury in its tissues might be quite healthy, might be harmful to man when ingested or vice versa.

A guide to the effects of organomercury compounds on man is provided by the World Health Organisation Maximum Permitted Limit of organomercury in fish that may be ingested by man. This regulation states that the mean daily consumption by man of organomercury shall not exceed 29 µg. Thus, if a man consumes $\frac{1}{4}$lb, i.e. 112 g, of fish daily then to be within the maximum limit, i.e. 29 µg organomercury per 112 g fish, the fish shall contain no more than 260 µg $kg^{-1}$ organomercury. A glance at the µg $kg^{-1}$ data in water creatures in

Table 2.15 reveals that the creatures that are most likely to cause harm to man are red and white tuna, halibut, trout, rainbow trout, pike, and swordfish and whale meat where, in all cases, the maximum reported organomercury contents might exceed 260 µg kg$^{-1}$.

### 3.1.3 ORGANOLEAD COMPOUNDS

The use of tetraalkyl leads as antiknock additives octane enhancers for automotive gasolines has now been reduced in many countries.

Another source of organolead compounds in the environment is the biological methylation of inorganic lead originating from industrial pollution. The biological methylation of inorganic lead to tetraalkyl lead compounds in lake sediments has been observed[21,26,44-47].

The high toxicity of tetraalkyl leads is attributed to their ability to undergo the following decomposition in the environment[48]:

$$R_4Pb \longrightarrow R_3Pb^2 \longrightarrow R_2Pb^2 \longrightarrow Pb^{2+}$$

The formation of trialkyl lead salts, probably associated with proteins, arising in tissues from rapid metabolic dealkylation of tetraalkyl lead compounds is of toxicological importance in evaluating exposure to tetraalkyl leads.

The possibility of biomethylation of lead or organolead ionic species by microorganisms (Wong et al.[47]; Schmidt and Huber[45]) reversing the decomposition mechanism given above, may add to environmental lead problems, although the area is presently disputed (Reisinger[21,49,26,48,50]).

The highly polar dialkyl and trialkyl lead compounds in particular have a high toxicity to mammals[51] and are formed as a result of the degradation of tetraalkyl lead in aqueous medium as shown above[48]. Tetramethyls and tetraethyl lead compounds are coincidentally more toxic than inorganic lead (1000 times[54]) or di- or tri-methyl or di- or tri-alkyl lead compounds[49]. In general organolead compounds are more toxic than inorganic lead compounds[52]. The toxicity of the alkylated lead compounds varies with the degree of alkylation, with tetraalkyl lead being the most toxic[49].

Toxicity of Organic Lead Compounds

| | | |
|---|---|---|
| Concentrations of Pb$^2$ (µg L$^{-1}$) which induces mortalities in 4–14 days | >100 | >1000 |
| Affected creatures | Annalids (adult) | Bivalve molluscs (adult) |
| | Bivalve Molluscs (larval) | Crustacea (adult) |
| | Crustacea (larval) | Echinoderms (adult) |
| | Gastropodes | Fish (adult) |

**Table 3.2.** Concentrations of organolead compounds found in natural waters and sediments and sea creatures.

| Total alkyl lead (excluding organic lead)* | River water ($\mu g\ L^{-1}$) | River sediments ($\mu g\ kg^{-1}$) | Coastal water ($\mu g\ L^{-1}$) | Coastal sediments ($\mu g\ kg^{-1}$) | In sea creatures ($\mu g\ kg^{-1}$) |
|---|---|---|---|---|---|
| | 50–530** | nd | <0.00001 | <0.00001 | 5090–18940 (carp) <1 (bass) 1760–5555 (small moult bass) 1520–1650 (pike) 7700–12600 (white sucker) <10–50 (crustacea) |

*Including $Me_4Pb$, $Me_3Pb$, $Me_3EtPb$, $Me_2PbEt_2$, $MePbEt_3$, $Et_4Pb$, $Me_3Pb^+$, $Me_2Pb^{2+}$, $Et_3Pb^+$, $Et_ePb^{2+}$
**Exceptional value due to gasoline spillage[53].

As stated above, tetramethyl and tetraethyl lead compounds are approximately one thousand times more toxic[54] than inorganic lead or di- or tri-methyl or di- or tri-ethyl lead compounds. The concentrations of tetraalkyl lead compounds in water which would cause mortalities are thus 1000 times less than those quoted above for inorganic lead, i.e. are reduced to $> 0.1\ \mu g\ L^{-1}$ in the case of adult annelids, larval bivalve molluscs, larval crustacea and gastropodes, or $1\ \mu g\ L^{-1}$ in the case of adult bivalve molluscs, crustacea, echinoderms and fish.

As seen in Table 3.2 the concentrations of organolead compounds occurring in coastal and, presumably, seawater are considerably below the $0.1–1\ \mu g\ L^{-1}$ range and, consequently, no adverse effects due to organolead are to be expected. However, the higher concentrations of organolead found in rivers (up to 530 $\mu g\ L^{-1}$ in an exceptional case where a tetraethyl lead spillage was known to have occurred) could cause mortalities in a wide range of creatures.

Reported concentrations of organolead found in freshwater creatures range from $< 1\ \mu g\ L^{-1}$ in bass to 18 940 $\mu g\ kg^{-1}$ in carp (Table 3.2). At the higher levels in tissues this represents a considerable accumulation of organolead compounds. Many of these creatures are at risk of ill health or mortality if the water in which they live contains more than $0.1–1.0\ \mu g\ L^{-1}$ of tetraalkyl lead. Also, at higher levels in the fish, people who eat the fish are at risk of adverse effects to his health.

Recently, a renewed interest in the speciation of lead in environmental samples has resulted from several diverse lines of investigation. Organolead compounds have been detected in cod, lobster, mackeral and flounder meal (up to 10 to 90% of the total lead burden) (Sirota and Uthe[55]), and in freshwater fish (Chau and Wong[56], Reamer et al.[57]). There is also evidence for the chemical (Reisinger[21]) and biological (Chau and Wong[56], Wong et al.[47]; Thayer[50]) alkylation of organolead salts or of lead (II) salts.

Speciation of alkyl lead compounds, including ionic, volatile and solvated forms has become important and in demand for environmental studies.

The toxicities of $R_3Pb^+$ and $R_2Pb^{2+}$ compounds are, amongst others, due to their abilities to inhibit oxidative phosphorylation (trialkyl lead compounds) and their affinity for thiols (dialkyl lead compounds) in animal body tissues and fluid.

Organically bound lead is a minor but important contribution to total lead intake by humans and animals. Alkyl lead salts such as trialkyl lead carbonates, nitrites and/or sulfates arising in tissues from rapid metabolic dealkylation of tetraalkyl lead compounds are of low toxicity.

Although organolead may make only a small contribution to the total lead intake in an organism, it has been demonstrated that trialkyl lead salts arising in tissues from the degradation of tetraalkyl leads are important in lead toxicity. The conversion of $R_4Pb$ to $R_3Pb^+$ occurs rapidly in liver homogenates from rats and rabbits. Acute toxicities of tetraalkyl lead and of trialkyl lead salts are similar and are at least an order of magnitude greater than dialkyl lead salts or inorganic lead salts. Relatively little is known either of the effect of chronic exposure to small amounts of such compounds or the levels of organic lead compounds, such as the tetraalkyl leads, in biological and food material.

The concentrations of inorganic lead in various types of water which can induce mortalities when various creatures are subject to short term (4–14 days) are listed in Table 3.1.

Fairly high concentrations of tetraalkyl lead (30 ppm) have been detected in mussels collected at a buoy near the S. S. Cavtat incident where a shipload of tetraethyl lead was sunk[55] in the Adriatic Sea. High organolead concentrations, mainly of tetraalkyl lead, were also found in mussels in other parts of Italian seas. The presence of tetraethyl lead in aquatic organisms may indicate that the alkyl lead compounds are not immediately metabolised by living organisms and may remain in their authentic forms in the living tissues for a long time[58]. The occurrence of tetraalkyl lead compounds in aquatic biota is highly significant because of the possibility of their incorporation into the food chain.

### 3.1.4 ORGANOTIN COMPOUNDS

These compounds have been the subject of environmental studies for two obvious reasons. Firstly is the increasing world wide use of inorganic and organotin compounds in many industrial, chemical and agricultural areas, very little being known about their environmental fate; secondly, there is a great difference in toxicity of various organotin compounds according to the variation of the organic moiety in the molecules.

Organotin compounds occur extensively in the ecosystem, being found in natural and marine waters, sea creatures and sediments. This originates from the use of these compounds in industry (PVC stabilisers) and agriculture (fungicides, biocides, bacteriocides) and in the marine environment (algicides and molluscicides[59-62]). Compounds such as triphenyl tin acetate and triphenyl tin hydroxide are used as molluscicides in antifoulant paint compositions for ships and harbour works.

Annual world production of organotin compounds was estimated to be 30 000 tons in 1983, most of it dioctyl tin maleate[60,63-65]. The toxicity and degradation in the environment depend strongly on the number and nature of the substituents[59,61,65]. Toxicity, environmental movement, and persistance of organotin compounds varies by several orders of magnitude depending on the number of carbon atoms present, chain length of alkyl groups and cyclic properties of the organotin compound. It is further complicated by structure (tetrahedral, trigonal, bipyrimidal or octahedral) and susceptibility to biodegradation.

Compounds with short organotin alkyl chains or phenyl substituents generally exhibit considerable toxicity towards both aquatic organisms and mammals. Alkyltins with small alkyl chains degrade slowly in the environment[66,67]. Phenyltin compounds are less stable and may, under certain conditions rapidly lose phenyl substituents[61]. Organotin compounds may accumulate in sediments and aquatic organisms[66].

There is a special interest in the biotic and abiotic methylation of tin compounds (Guard et al.[68]) and the fate of some industrial organotins in the aquatic ecosystems. One possible route is the dealkylation of the trialkyltin species eventually to Sn(IV) and the microbial methylation of Sn(IV) to the various methyltin species. Increasing methylation concentrations with increasing anthropogenic tin influxes has been noted in the Chesapeake Bay (Jackson et al.[69]).

Trialkyltin species originating as organotin pollutants can dealkylate eventually to inorganic tin in the aqueous ecosystem.

$$R_3Sn^+ \longrightarrow R_2Sn^{2+} \longrightarrow RSn^{3+} \longrightarrow Sn^{4+}$$

Rapsomankis and Weber[70] examined the environmental implications of the methylation of tin(II) and methyltin(IV) ions in aqueous samples in the presence of manganese dioxide.

Their studies were carried out with particular reference to the mechanisms involved and the role of a dimethylcobalt complex carbanion donor, the carbocation donor iodomethane, and the oxidising agent manganese dioxide. The yields of the various methyltin ions were estimated: some preliminary results were also presented on the further methylation of monomethyltin, dimethyltin, and trimethyltin, which indicated that the presence of a naturally occurring donor such as methylcobalamin would result in the formation of volatile tetramethytin and this would account for the global occurrence of methyltin compounds.

Van Nguyen et al.[71] carried out an investigation of the fate in an aqueous environment of three organotin compounds (triphenyltin acetate, triphenyltin hydroxide and triphenyltin chloride) used in antifoulant paint compositions. The organotin compounds were leached from paint panels by shaking with distilled water for up to 2 weeks at room temperature, and the water and undissolved residues were then examined by infrared spectroscopy and thin-layer chromatography. The results suggested that the organotin compounds ionised in aqueous media; a simple model was developed to explain the process.

Methyltin species are ubiquitous in natural waters, although their concentration is usually low (less than $1 \mu g L^{-1}$) in waters relatively unimpacted by anthropogenic activity[72,73]. Mono and dimethyltin are the dominant species[72-74], suggesting that methyltins, like methylmercury species, arise via stepwise methylation of the inorganic metal[75]. Not only are sediment slurries capable of methylating added inorganic tin[76], but concentrations of methyltin species increase with estuarine surface-to-volume ratios[72]. Thus methyltin in aquatic environments is likely to occur in sediments.

Organotin compounds can be produced in the ecosystem by biologically induced methylation of inorganic tin compounds. Measurements of sediment methyltin concentrations show monomethyltin to be the dominant species in anoxic sediments whilst trimethyltin is found in highest concentrations in oxic (anaerobic) sediments[77]. This suggests that tin methylation probably occurs in anaerobic sediments, whilst degradation of higher molecular weight organotins such as tributyltin, an antifoulant agent, occurs in oxygenated environments. In studies of inorganic tin methylation it has been confirmed that biomethylation occurs preferentially in anaerobic estuarine sediments[78,79]. Methyltin compounds were produced to a maximum level of about 2 µg kg$^{-1}$ (dry weight) of sediment in 21 days. Concentrations of mono-, di- and tri-alkyl tin compounds found in Baltimore Harbour sediments averaged at 8, 1 and 0.3 µg kg$^{-1}$ dry weight of sediment respectively whilst sediments taken in a relatively unpolluted area had much lower organotin content (1.0, 0.1 and 0.01 µg kg$^{-1}$ respectively)[78,79].

Several investigators have reported ng–µg L$^{-1}$ concentrations of organotin compounds in both freshwater and marine samples. Inorganic tin, methyltins and butyltins have been detected in marine and freshwater environmental samples[80-83]. Both organotins and inorganic tins were reported to be highly concentrated by factors of up to 10$^4$ in the surface microlayer relative to subsurface water[84,85] Methylation of tin compounds by biotic as well as abiotic processes have been proposed[86,87].

Possible antropogenic sources of organotins have recently been suggested. Both polyvinylchloride and chlorinated polyvinylchloride have been shown to leach methyltin and bibutyltin compounds, respectively, into the environment[88]. Monobutyltin has been measured in marine sediments collected in areas associated with boating and shipping. Butyltin was not detected in areas free of exposure to maritime activity[89]. The use of organotin antifouling coatings in particular has stimulated interest in their environmental impact.

Zischke et al.[90] have determined 96h LC$_{50}$ values of tributyltin compounds to mysids (*Mysidopes bahia*). The age of the fish was an important factor in determining the sensitivity of juveniles to tributyltin compounds.

In chronic toxicity tests[91,92] carried out in the Chesapeake Bay area in which biota were exposed to tributyltin the survival of *Gammarus SP* was unaffected by 24 h exposure to concentrations up to 0.58 µg L$^{-1}$, although body weight was reduced by 64% relative to controls. Survival of *Brevoortia tyrannus* and larval *Menidia beryllina* was unaffected by 28 days exposure to concentrations of tributyltin up to 0.49 µg L$^{-1}$. Extensive variation between individuals made it difficult to discern treatment related histological changes in *Brevootra tyrannus*.

In *Mendia beryllina*, growth was reduced by 20–22% following exposure to 0.09 or 0.49 µg $L^{-1}$ tributyltin, but various morphometric measurements revealed no treatment related effects. It was considered unlikely that mean environmental concentrations of tributyltin in Chesapeake Bay marinas would cause direct mortality to the 3 test species after 4 weeks exposure, although sublethal effects could occur.

A 24 h $LC_{50}$ value of 1–3 µg $L^{-1}$ has been reported for adult rainbow trout. The concentrations of tributyltin found in surface microlayers of natural waters were in the range 1.9–473 µg $L^{-1}$. Consequently, surface swimming rainbow trout could be at risk.

The occurrence and concentrations of organotin compounds in the tissues of scallops (*Pecten maximus*), flame shells (*Lima hiams*)[93], polychaetes, snails and bivalves[94], and mussels (*Mylilus edulis*) and oysters (*Crassostrea virginica*)[95] has been studied. Scallops, mussel and flame shells populations are adversely affected by organotin compounds[93]. High concentrations of tributyltin have been found in polychaetes, snails and bivalves living in marinas containing 2–646 ng $L^{-1}$ tributyltin[94] i.e. above the Environmental Quality Target for tributyltin of 20 ng $L^{-1}$. San Diego Bay mussels exposed to 0.7 µg $L^{1-}$ organotin for 60 days sustained a 50% mortality in the case of mussels and a decline in condition in the case of oysters[95]. Various tissues in the organisms showed tin uptake within 0–30 days.

Roberts[96] demonstrated that acute toxicity with tributyltin chloride (using glacial acetic acid as a carrier) yielded 48h $LC_{50}$ values of 1.30 and 3.96 µg $L^{-1}$ in *Crassostrea virginica* embryos and straight hinge stage larvae, respectively, and 1.13 and 1.65 µg $L^{-1}$ in *Mercenaria mercenaria* embryos and larvae, respectively. The 24 $LC_{50}$ (both species) were greater than 1.3 µg $L^{-1}$ in embryos and greater than 4.2 µg $L^{-1}$ in larvae. In the one experiment which used acetone as a carrier for tributyltin chloride, the 38 h $LC_{50}$ for *Crassostrea virginica* embryos was 0.71 µg $L^{-1}$. Evidence suggested that tributyltin doses below the 48 h $LC_{50}$ delayed clam embryo development, although resultant larvae were not abnormal. In addition to slightly delaying oyster embryo development, tributyltin (0.77 µg $L^{-1}$ and above) caused abnormal shell development (flattened rather than convex shells) in resultant larvae. Comparing these results with those obtained in tests with *Crassostrea Gigas, Mytilus edulis* and *Mytilus galloprovincialis* indicated that embryos and larvae of the 5 species of bivalve mollusc had similar sensitivities to tributyltin and that tolerance increased slightly with increasing larval age.

Weis et al.[97] showed that exposure of adult filder crabs to tributyltin concentrations as low as 0.5 µg $L^{-1}$ retarded limb regeneration and ecdysis, and produced morphological abnormalities in regenerated limbs. Deformities included backward curling of the dactyl of the claw, curling and stunting of walking legs and a reduction in the number of setae. In two of the trials, males appeared more sensitive to tributyltin than females, whereas the reverse was true in the third trial. These differences might have been related to seasonal reproductive activity.

It was also observed that the presence of sediment greatly reduced the effects of tributyltin.

In a 13 month experiment conducted by His and Roberts[98] in Arachon Bay, oysters were cultivated in trays with wooden sides which had been painted with organotin antifouling paint (tributyltin) or with copper oxide antifouling paint (Renaudin La Precieuse). The organotin paint adversely affected oyster weight, length and width as compared with controls (in unpainted trays), but did not affect shellheight. Shell density (reflective of shell calcification) and dry condition factors were markedly decreased. Embryonic and larval viability was unaffected, but larval growth was slightly impaired. Copper oxide paint has no effect on oyster growth or shell calcification. Dry condition factors were reduced, but to a lesser extent than was observed with organotin. Embryonic and larval development were unaffected by the copper oxide paint.

Bailey and Davies[99] showed that the degree of imposex in populations of dogwhelks from 30 sites in Sullom Voe indicated tributyltin contamination of coastal waters close to an oil terminal where large tankers docked. Imposex is quantified by calculating the relative penis size index, ranged from 34.03 to 81.37% within the Voe and up to 10 km from the terminal suggesting considerable transport and little dilution of the tributyltin. The vas deferens sequence index was greater than 4.0 in high proportions of females from all populations in the Voe indicating significant impairment of reproductive ability.

Tin was detected in dogwhelk samples from inside the Voe at concentrations up to $0.1$ mg kg$^{-1}$ and low concentrations (less than $0.03$ mg kg$^{-1}$ wet weight) were found in the edible tissue of queen scallops inside the Voe but tin was rarely detected in commercial shellfish outside the Voe. Tributyltin was detectable (2 ng L$^{-1}$ as tin) in only 2 of the 7 sampling sites around the terminal area. The sources of tributyltin contamination were antifoulants on tankers and towing vessels and from previous use on navigational buoys and harbour craft.

Goodman et al.[100] used the estuarine fish *Mysidopsis bahia* as the test organism in several procedures for toxicity testing, but the recommended age of the test animals varied with different methods. To obtain information on the effect of age on the acute sensitivity of this species to toxic compounds, fish aged less than 1,5 and 10 days old at the start of the experiment were exposed to 3 toxic compounds. The 96 h LC$_{50}$ values of the experimental data indicated that age was not an important factor in the sensitivity of juvenile *Musidopsis bahia* to the compounds tested.

The adverse effects of organotin compounds in water at the $0.5$ g L$^{-1}$ level on fish and creatures other than fish in saline waters are summarised below:

LC$_{50}$ values obtained for organotin compounds are extremely low, confirming the high toxicity of these compounds towards water creatures. Reported values for 1 and 2 day LC$_{50}$s are in the range 1–4 µg L$^{-1}$ as seen in Table 3.3

|  | Non saline waters | Saline waters | |
|---|---|---|---|
|  | Tributyltin (0.5 µg $L^{-1}$) | | |
| Fiddler crabs (Uca pugilator) | Retarded limb regeneration[97] morpholical abnormalities i.e. Regenerated limbs | Gammarus GP | Reduced bodyweight[91] |
|  |  | Brevoortia tyrannus and larval Menidia beryllina | Reduced growth rate |
| Bivalve molluscs (Crassotroa virginica and Mercenaria Merconaria) | Tributyltin Acute toxicity to embryos and larvae, delayed clam embryo development[96] (below $LC_{50}$ value) | Oyster Crassostrea gigas | Weight, length, width adversely affected[98] |
|  | Tributyltin abnormal shell development at greater than 0.77 µg $L^{-1}$ tributyltin in water. | | |

Table 3.3. $LC_{50}$ values obtained for organotin compounds.

|  | Species | Type of water | $LC_{50}$ | Duration of $LC_{50}$ test (days) | Ref. |
|---|---|---|---|---|---|
| Organotin Compounds | Rainbow trout (Salmo gairdneri) | Non saline | 1.3 µg $L^{-1}$ | 1 | 104 |
| Organotin Compounds (tributyltin) | Bivalve mollusc (Crassostrea virginica) | Non saline | 1.3 µg $L^{-1}$ embryo 3.96 µg $L^{-1}$ | 2 | 96 |
| Organotin Compounds (tributyltin) | Bivalve mollusc (Mercenaria mercenaria) | Non-Saline | 1.13 µg $L^{-1}$ embryo 1.65 µg $L^{-1}$ larvae | 2 | 96 |
| Organotin Compounds (tributyltin) | Bivalve mollusc (Crassostrea virginica) | Non-Saline | >1.3 µg $L^{-1}$ embryo 4.2 µg $L^{-1}$ larvae | 1 1 | 96 |
| Organotin Compounds (tributyltin) | Bivalve mollusc (Mercenaria mercenaria) | Non-Saline | >1.3 µg $L^{-1}$ larvae 4.2 µg $L^{-1}$ larvae | 1 1 | 96 |

## 3.1.5 ORGANOARSENIC COMPOUNDS

Large amounts of arsenic enter the environment each year due to the use of arsenic compounds in agriculture and industry as pesticides, herbicides and wood and crop preservatives. The main form in which it is introduced is as the inorganic arsenic (arsenite and arsenate). Arsenic is easily transformed between its inorganic and organic forms by biological and chemical action. As the toxicity and biological activity of the different species vary considerably, information on the chemical form, i.e. speciation, is of great importance in environmental studies. Thus, organoarsenic in the trivalent state is very toxic to most organisms but in the pentavalent state it may be an essential nutrient for crustacea and seems to be innoxious to mammals.

The biological methylation of inorganic arsenic by microorganisms such as molds and bacteria present in sediments, sludges and muds has been established although there is no unequivocal evidence of the proposed pathways[5,6,11,20,23-25].

Organoarsenic species are known to vary considerably in their toxicity to humans and animals as discussed above[101]. Large fluxes of inorganic arsenic into the aquatic environment can be traced to geothermal systems[102], base metal smelter emissions and localised arsenite treatments for aquatic weed control. The methylated arsenicals have entered the environment either directly as pesticides or by biological transformation of the inorganic species[103,47].

Organoarsenical pesticides such as sodium methylarsenate and dimethyl arsenic acid are used in agriculture as herbicides and fungicides. It is possible that these arsenicals enter soil, plant and consequently humans. On the other hand, arsenic is a ubiquitous element on the earth, and the presence of inorganic arsenic and several methylated forms of arsenic as monomethyl, dimethyl and trimethylarsenic compounds in the environment has been well documented (Braman[105]). The occurrence of biomethylation of arsenic in microorganisms (Cox[106]), soil (Lasko and Peoples[75]) has also been demonstrated. Therefore, further investigation of the fate of arsenicas in the physical environment and living organisms requires analytical methods for the complete speciation of these arsenicals.

It has been shown that arsenic is incorporated into both marine and freshwater organisms in the form of both water soluble and lipid soluble arsenic compounds (Chapman[107]). Recent studies to identify the chemical forms of these arsenic compounds have shown the presence of arsenite (As-III), arsenate (As-V), methylarsonic acid, dimethylarsinic acid and arsenobetaine (Andreae[103]). Methylated arsenical also appear in the urine and plasma of mammals, including man, by biotransformation of inorganic arsenic compounds (Lasko and Peoples[75]). Several methods have been devised to characterise these arsenicals.

Inorganic arsenic has similar toxic properties to lead and mercury as regards its ability to bond to sulphur and to inhibit enzyme action such as pyruvate dehydrogenase. It tends to concentrate in the liver, kidneys and lungs of animals exposed to it. Exposure of fish to 4000 µg L$^{-1}$ of inorganic arsenic for 39 days causes adverse effects such as reduced rate of growth. Marine crustacea are the most sensitive to arsenic and annelids and insect larvae the least sensitive. Marine organisms contain widely varying concentrations of arsenic in different chemical forms, both organic and inorganic. A knowledge of these chemical forms and their bio-availability is important. Crab is susceptible to arsenic at its larval life stage. In the case of crustacea toxicity decreases in the order arsenate, organic and arsenite. Fish are less susceptible to arsenic than invertebrates.

Cockell and Hilten[108] fed juvenile raniblow trout for 8 weeks on semi-purified diets containing arsenic trioxide (at-180–1477 µg arsenic $g^{-1}$ diet), disodium arsenate (137–1053 µg arsenic $g^{-1}$), dimethylarsinic acid (193–1503 µg arsenic $g^{-1}$) or arsanalic acid (193–1503 µg arsenic $g^{-1}$). Growth, food consumption and feeding behaviour were adversely affected by all dietary concentrations of inorganic arsenicals, but were unaffected by diets containing organic arsenicals. In all treated trout, carcass arsenic concentrations were related to dietary arsenic concentration and to dietary arsenic exposure rate (mg arsenic $kg^{-1}$ body weight day). Carcass arsenic concentrations were highest in trout fed inorganic arsenicals, with disodium arsenate and arsenic trioxide yielding highest carcass arsenic concentrations at low and high exposure levels, respectively.

The UK Arsenic in Food Regulations (1959) state that foodstuffs must not contain more than 1000 µg $kg^{-1}$ of total arsenic. Fish and edible seaweeds are made an exception as their arsenic contents are normally in excess of 1000 µg $kg^{-1}$. The UK total diet survey suggests that at least 75% of the total amount of arsenic ingested by man originates from fish and shellfish. It is accepted that the arsenic in these foods is mainly organically bound i.e. in its least toxic form. If the levels of total inorganic arsenic approaches 1000 µg $kg^{-1}$ then a knowledge of the proportion of the least toxic arsenite to the most toxic arsenate is important.

## 3.2 TOXIC LEVELS IN THE HUMAN DUCT

To carry out calculations of the toxic hazard present by metals and organometallic compounds to man it is necessary to know the range of concentrations of each toxicant found in representative samples of foods and also to know the average daily intake of these foods by man. It is then possible to calculate the weight of each toxicant ingested daily. This data, together with toxicological data for each toxicant, enables the risk to health of humans to be evaluated, and decisions to be made as to whether or not the foodstuffs are acceptable for human consumption. In the UK it is standard practice to assume that each person, on average, eats 1.5 kg of food per day. In other countries higher consumption rates used e.g. Canada 1.8 kg per person per day, and in the U.S.A. 3.78 kg per person per day. The UK rate of food consumption is assumed in the following calculations.

### 3.2.1 INORGANIC ELEMENTS

The data in Table 3.4(A) lists ranges of concentrations of some metal ions found in twelve different types of food, also in drinking water. These metals were selected because they are all capable of existing also in organometallic forms. It is assumed that the human consumes, on average, 1/12th of 1.5 kg, i.e. 0.125 kg of each food per day. On this basis the ranges of weight of each toxicant consumed daily per person can be calculated (Table 3.4).

It should be noted that these values are minimum because analytical data is not available for some elements in some foods (note the nd comments).

A further source of intake of metals by humans is drinking water. From the available analytical data on drinking water ($M$ to $M^{-1}$ µg L$^{-1}$) and assuming a daily average consumption per person of four litres, it is possible to calculate the range of weights ($4M$ to $4M^{-1}$ µg) of these ions consumed daily per person (Table 3.4B). The total intake of metals from foods and water in (W µg per person per day) from a daily diet consisting of 1.5 kg food and 4 L water can then be calculated.

In Table 3.5 these $W$ values are compared with UK Government statistics for daily intakes of elements from an assumed standard diet of food ($W^1$ µg per person per day). These values are appreciably lower than those obtained by actual analysis of foodstuffs. In Table 3.5 the $W$ and $W^1$ values are compared with World Health Organisation data for maximum tolerable daily consumptions per person of these elements from food and water.

Regarding mercury estimates ($W$ and $W^1$) of the daily intake via foods (5–19.3 µg per person/day) are between lower and higher than the WHO maximum limit of 14 µg per person/day for toxicological safety.

Regarding lead estimates of the daily intake via foods ($W$ and $W^1$ 170–4015 µg per person per day) indicate that the WHO maximum for toxicological safety of 430 µg per person per day can sometimes be exceeded (170–4015 µg per person per day) especially if the diet consists of highly contaminated potatoes, green vegetables, meat, fish or crustacea (see Table 3.4).

Sometimes higher than safe concentrations of elements in the solid and liquid diet can be attributed to particular items of the diet. Thus, as shown in Table 3.4, crustacea can contain up to 2800 µg kg$^{-1}$ of lead, liver can contain up to 11 700 µg kg$^{-1}$ manganese, flours can contain 8700–20 000 µg kg$^{-1}$ manganese and spinach and kale can be particularly rich in manganese (173 000 µg kg$^{-1}$) and lead (2900–3200 µg kg$^{-1}$).

Assuming, for example, that a consumer eats 125 kg of crab containing lead, then to meet the WHO requirement for lead in his diet (i.e. 430 µg lead per day from all sources, say 140 µg from crab only) then the crab would have to contain a maximum of 1100 µg kg$^{-1}$ of lead. The minimum quoted value for lead in crustacea is 500 µg kg$^{-1}$ and the maximum 2800 µg kg (Table 3.4). So it can be seen that some, at least, of the crab caught are acceptable for human consumption. Conversely, some are not.

Due to the lack of analytical and/or toxicological data, realistic comments cannot be made on the situation regarding other inorganic elements. Yet, as shown in Table 3.4 quite considerable quantities of these elements can be ingested by man in his diet.

TOXIC EFFECTS OF ORGANOMETALLIC COMPOUNDS

**Table 3.4.** Ranges of inorganic metal contents found in foods and drinks water (A) and daily intakes by humans (B).

| A Element | Cereals ($\mu g\ kg^{-1}$) | | | | Vegetables and Fruit ($\mu g\ kg^{-1}$) | | | | Meat ($\mu g\ kg^{-1}$) |
|---|---|---|---|---|---|---|---|---|---|
| | (1) Corn | (2) Wheat flour | (3) Rice flour | (4) Potatoes | (5) Kale | (6) Spinach | (7) Fruit | | (8) Liver |
| Mercury | nd | nd | nd | 155 | nd | nd | nd | | nd |
| Lead | 260–610 | 30–160 | nd | 30–1200 | 1600–2900 | 440–3200 | nd | | 250–1130 |
| Copper | 100–7500 | 1300–2000 | 1300–5200 | 2500–3100 | 4000–6000 | 5000–13 900 | 200–300 | | 2000–3100 |
| Nickel | 2800–3600 | 10–60 | nd | nd | 800–1100 | 2600–7500 | nd | | 20–120 |
| Manganese | nd | 5800–8500 | 19 900 | 4110–5670 | 14 100–16 900 | 156 000–173 000 | 530–660 | | 9700–11 700 |
| Selenium | 5–530 | nd | 570–1870 | nd | 133–138 | 2–780 | 2–42 | | nd |
| Antimony | nd | <2 | nd | nd | nd | 10–40 | nd | | nd |

| B | Metal Content ($\mu g$) of 0.125 kg of each of the food items 1–12 and of 4L of drinking water | | | | | | | | |
|---|---|---|---|---|---|---|---|---|---|
| Mercury | nd | nd | nd | 19.3 | nd | nd | nd | | nd |
| Lead | 32.5–76 | 3.7–20 | nd | 0.37–150 | 200–362 | 55–400 | nd | | 31.2–141.2 |
| Copper | 12.5–950 | 162–250 | 162–650 | 312–387 | 500–750 | 1000–1737 | 25–37.5 | | 250–387 |
| Nickel | 350–450 | 1.2–7.5 | nd | nd | 100–137 | 325–937 | nd | | 2.5–15 |
| Manganese | nd | 725–1062 | 2487 | 513–709 | 1762–2112 | 19 500–21 620 | 66.2–82.5 | | 1212–1462 |
| Selenium | 0.62–66.2 | nd | 108.7–234 | nd | 16.6–17.2 | 0.25–97.5 | 0.25–5.2 | | nd |
| Antimony | nd | <0.25 | nd | nd | nd | 1.2–5.0 | nd | | nd |

nd—Not Determined.

**Table 3.4.** (*continued*)

| A Element | (9) Milk powder (μg kg$^{-1}$) | Dairy Products (μg kg$^{-1}$) (10) Cheese (μg kg$^{-1}$) | (11) Fish (μg kg$^{-1}$) | (12) Crustacea (μg kg$^{-1}$) | (13) Drinking water (μg L$^{-1}$) |
|---|---|---|---|---|---|
| Mercury | nd | nd | nd | nd | 0.001–0.01 |
| Lead | 20–630 | nd | 100–1400 | 500–2800 | <0.6–565 |
| Copper | 400–1000 | 20 | 400–2200 | 700–2600 | <3–945 |
| Nickel | 30–400 | nd | 150–200 | nd | 0.19–6.1 |
| Manganese | 200–760 | nd | 200–1600 | nd | 1.9–20.5 |
| Selenium | 3–15 800 | nd | 190–550 | 700–6700 | <10 |
| Antimony | nd | nd | nd | 70–400 | <13 |

| B | | | | | Total solid food Plus Water W (μg) |
|---|---|---|---|---|---|
| Mercury | nd | nd | nd | nd | 19.3 |
| Lead | 2.5–78.7 | nd | 12.5–175 | 62.5–350 | 402.7–4015 |
| Copper | 50–125 | 2.5 | 50–275 | 87.5–325 | 2625.5–9656 |
| Nickel | 3.7–50 | nd | 18.7–25 | nd | 801.9–1646 |
| Manganese | 25–95 | nd | 25–200 | nd | 26 257–29 829 |
| Selenium | 0.37–1975 | nd | 23.7–68.7 | 87.5–837 | 278.0–3341 |
| Antimony | nd | nd | nd | 8.7–50 | 9.9–55.0 |

# TOXIC EFFECTS OF ORGANOMETALLIC COMPOUNDS

**Table 3.5.** Comparison of calculated ($W$) and recommended maximum ($W^1$) daily intakes of elements.

| | From Table 3.4($W$) Daily intake (µg) per person per day from solid foods* and 4 L drinking water | Mean intake (µg) per person per day from solid foods UK Official Data $W^1$ | From solid foods** µg per person per day | WHO Limits for Toxicological Safety | | |
|---|---|---|---|---|---|---|
| | | | | From 4 L drinking water µg/person/day | From solid foods plus drinking water µg/person/day | µg/person/week |
| Mercury | 19.3 | 5–10 | 14 i.e. 98 µgHg/week | 2(WHO) 8 (EPA) | 16 | 112 |
| Lead | 402.7–4015 | 170–280 | 430 i.e. 3010 µgPb/week | — | 22 430 | 154 3010 |
| Copper | 2625–9656 | 2000–5000 | — | 200 | 200 | 1400 |
| Nickel | 801.9–1646 | — | — | — | — | — |
| Manganese | 26 257–29 829 | 1600–10000 | 2000–3000 | — | 2000–3000 | 14 000–21 000 |
| Selenium | 278.0–3341 | 200 | — | — | — | — |
| Antimony | 9.9–55.0 | — | — | — | — | — |

*See Table 3.4(B) cereals, vegetables and fruit, meat, dairy products, fish and crustacea.
WHO World Health Organisation.
EPA Environmental Protection Agency.
**Canned foods, beverages, cereals, vegetables, fruit, meat, poultry, dairy products, fish, crustacea, yeast, spices, nuts and pulses.

## 3.2.2 ORGANOMETALLIC COMPOUNDS

Toxicological data for organometallic compounds is very sparse and so it is difficult or impossible to evaluate the risk to man involved in consuming these substances. Clearly, this is an area where detailed toxicological studies need to be carried out and, until they are, the only safe compromise is to limit severely the amount of these substances released into the ecosystem and to exert sensible controls on the amounts eaten of foodstuffs containing these substances.

In the U.K. Regulations exist concerning the levels of various toxicants in fish and shellfish. These are based on the maximum acceptable consumptions of these foods during one week. It is stated that the 90th percentile consumption of fish should not exceed 0.79 kg per person per week and of shellfish 0.26 kg per person per week.

Within these weekly food consumption limits it is seen in Table 3.6 that depending on contaminant levels it is possible for the consumer to ingest quite considerable quantities of organometallic compounds and the corresponding inorganic metals in a weekly diet which includes a maximum 0.79 kg fish and 0.26 kg shellfish. Additional dietary intakes will occur, of course, from other foodstuffs and potable water.

Thus in the case of inorganic lead, based on the analysis of fish and shellfish and assuming weekly consumptions, of respectively 0.79 kg and 0.26 kg, it is seen that between 107.3 and 1802.4 µg of inorganic lead is consumed. Both are within the WHO maximum limit for toxicological safety of 3010 µg per person per week. However, additional dietary inputs of lead from other foodstuffs must be taken into account in the final analysis. Based on alkyl lead contents in fish and shellfish and assuming again weekly consumptions of 0.79 and 0.26 kg it is seen that the levels of alkyl lead ingested per week (3.39–1494.4 µg) are similar to those of inorganic lead (107.3–1802.4 µg). However, the toxicity of alkyl lead is considerably higher than that of inorganic lead and consequently, to meet toxicological safety requirements, only an extremely low alkyl lead content can be allowed in fish and shellfish.

The situation regarding inorganic mercury and organomercury is similar to that of lead. Only when mercury levels are at the lower end of the observed concentration ranges found in fish and shellfish will toxicological safety requirements be met. The EC mercury in fish regulations state that fish shall not contain more than 30 µg kg$^{-1}$ of total inorganic plus organometallic mercury. The UK Arsenic in Food Regulations (1959) state that foodstuffs shall not contain more than 1000 µg kg$^{-1}$ of total arsenic. The observed arsenic contents of fish (1100–2900 µg kg$^{-1}$) and crustacea (20–180 µg kg$^{-1}$) (see Table 3.6) would imply that the consumption by humans of 0.79 kg fish and 0.26 kg crustacea weekly would lead to the consumption of

$$1100 \times 0.79 + 20 \times 0.26 \text{ to } 2900 \times 0.79 + 180 \times 0.26 = 874 \text{ to } 2338 \text{ µg}$$

arsenic equivalent to 872 to 2338 µg arsenic per 1.05 kg food i.e. 832 to 2226 µg arsenic kg$^{-1}$ plus any other arsenic ingested from other foods. Thus, at the higher end of the observed range (2226 µg kg$^{-1}$ total arsenic) the seafood would not pass

**Table 3.6.** Weights of organometallic and corresponding metals consumed per week by a human consuming 0.79 kg fish and 0.26 kg shellfish.

| Compound | Concentration range, $F(\mu g\ kg^{-1})$ (see Table 2.16) | Weight (μg) of toxicant consumed per week per person by eating 0.79 kg fish i.e. $0.79 \times F$ kg | Concentration range $C(\mu g\ kg^{-1})$ (see Table 2.18) | Weight (μg) of toxicant consumed per person per week eating 0.26 kg shellfish i.e. $0.26 \times C$ | Total weight of toxicant consumed per person per week $(0.79F + 0.26C)$ kg | WHO maximum limits for toxicological safety μg per person per week (see Table 3.4) |
|---|---|---|---|---|---|---|
| Alkyl lead | 1–18 900 | 0.79–14 931 | <10–50 | <2.6–13.0 | 3.39–14 944 | — |
| Inorganic lead | 120–1360 | 94.8–1074.4 | 48–2800 | 12.5–728 | 107.3–1802.4 | 3010 |
| Alkyl mercury | 50–8400 | 39.5–6636 | — | — | 39.5–6636 | 203[a] |
| Inorganic Mercury | 90–2400 | 71.1–1896 | 20–310 | 52–80.6 | 76.3–1976.6 | 98[a] |
| Alkylarsenic | — | — | 1400–23 000 | 364–5980 | 364–5980 | — |
| Inorganic Arsenic | 1100–2900 | 869–2291 | 20–180 | 5.2–46.8 | 874.2–2337.8 | — |

(a) WHO states maximum of 0.29 μg ethyl mercury per person per day, i.e. 203 μg per person per week and 43 μg total mercury per person per day i.e. 301 μg per person per week. Therefore, inorganic mercury is 14 μg per person per day i.e. 98 μg per person per week.

the UK regulations of 1000 µg kg$^{-1}$, whereas at the lower end of the observed range (832 µg kg$^{-1}$) it would.

Because about 75% of the arsenic ingested in the human diet originates from fish and shellfish the UK Regulations permit an exception in which the arsenic content can exceed 1000 µg kg$^{-1}$. This is because, it is claimed, the arsenic is present mainly in the less toxic organic form. If, however, the arsenic is known to be present in the more toxic inorganic form then the limit of 1000 µg kg$^{-1}$ is imposed.

## REFERENCES

1. Uthe J.F., Armstrong F.A.J., Tam K.J. *Ass. Offi. Anal. Chem.* **54** 866 (1971)
2. Backmann K. *Talanta* **29** 1 (1982)
3. Ridley W.P., Diziker L.J., Wood J.M. *Science* **197** 329 (1977)
4. Imura N., Pan E.S.S., Kim K.N.J., Ukita T.K.T. *Science* **172** 1248 (1971)
5. Wood J.M., Rosen C.G., Kennedy S.F. *Nature (London)* **220** 173 (1968)
6. Jensen S., Jernelov A. *Nature (London)* **223** 753 (1969)
7. Krenkel P.A. International Critical Review of Environmental Control **3** 303 (1973)
8. Robinson S., Scott W.B. A Selected Biography on Mercury in the Environment with Subject Listing, Life Science, Miscellaneous Publication, Royal Ontario Museum, P54 (1974)
9. Uthe J.F., Armstrong F.A.J. *Toxicological Environmental Chemistry Review* **2** 45 (1974)
10. D'Itri P.A., D'Itri F.M. Environmental Management **2** 3 (1978)
11. Wood J.M. *Science* **183** 1049 (1974)
12. Furunkawa K., Tanomura K. *Agricultural and Biological Chemistry* **35** 604 (1971)
13. Furunkawa K., Tanomura K. *Agricultural and Biological Chemistry* **36** 217 (1972)
14. Furunkawa K., Tanomura K. *Agricultural and Biological Chemistry* **36** 244 (1972)
15. Brinkmann F.E., Iverson W.P. In *Marine Chemistry in the Coastal Environment* (Editor T. Church) American Chemical Society. Synposium 18. American Chemical Society, Washington DC (1975)
16. Fogerstron T., Jernelov A. *Water Research* **5** 121 (1971)
17. Yamada M., Tanomura K. *Journal of Fermentation Technology* **50** 159 (1972)
18. Yamada M., Tanomura K. *Journal Fermentation Technology* **50** 893 (1972)
19. Yamada M., Tanomura K. *Journal of Fermentation Technology* **50** 901 (1972)
20. Van Hall C.E., Stenger V.A. *Analytical Chemistry* **39** 502 (1967)
21. Reisinger K., Stoeppler M., Nurnberg H.O. *Nature (London)* **291** 228 (1981)
22. Challenger F. *Chemical Reviews* **36** 315 (1975)
23. Vonk J.W., Sijperstein A.K. *Antorie van Leernwenhoek* **39** 505 (1973)
24. McBride B.C., Merilees H., Cullen W.R., Picket W. ACS Synposium Series **82** 94 (1978)
25. Landner L. *Nature (London)* **230** 452 (1971)
26. Wong P.T.S., Chanuy K., Luxcn P.L. *Nature (London)* **253** 263 (1975)
27. Von Endt D.W., Kearney P.C., Kaufman D.D. *J. Agricultural and Food Chemistry* **16** 17 (1968)
28. Westoo G., Rydalu M. *Var Foeda* **21** 20 (1969)
29. Westoo G., Rydalu V. *Var Foeda* **21** 138 (1969)
30. Westoo G. *Acta Scand Sc. Anal.* **20** 2131 (1968)
31. Westoo G. *Acta Scand Sc. Anal.* **21** 1790 (1967)
32. Westoo G. *Acta Scand Sc. Anal.* **22** 2277 (1968)

33. Jones P., Nickless P. *Analyst (London)* **103** 1121 (1978)
34. Collett D.L., Fleming J.E., Taylor G.E. *Analyst (London)* **105** 897 (1980)
35. Bache C.A., Lisk D.J. *Analytical Chemistry* **45** 950 (1971)
36. Vostal J. *Mercury in the Environment*, CRC Press, Cleveland, Ohio (1972)
37. Ram R.N., Sathyanesan A.G. *Environmental Pollution* **47** 135 (1987)
38. Kirubagarun R., Joy P. *Ecotoxicology and Environmental Safety* **15** 171 (1988)
39. Czuba M., Seagull R.W., Tran H., Cloutier L. Ecotoxicology and Environmental Safety **14** 64 (1987)
40. Davies I.M., Graham W.C., Pirie S.M. *Marine Chemistry* **7** 11 (1979)
41. Kienmeij A.M., Kloosterboer J.G. Analytical Chemistry **48** 575 (1976)
42. Minagawa K., Takizawa A., Kifune I. *Anal. Chem. Acta.* **115** 103 (1980)
43. Monoaro S., Cause B.S., Kirkbright G.C. *Water Research* **13** 503 (1979)
44. Jarvis A.W.P., Markell R.N., Potter H.R. *Nature (London)* **255** 217 (1975)
45. Schmidt U., Huber F. *Nature (London)* **259** 159 (1976)
46. Dumas J.P., LeRoy Pazdernik S., Bellonik D., Bauchard D., Vaill Ancourt G. Pnoc. 12th Canadian Symposium on Water Pollution Research **12** 91 (1977)
47. Wong P.T.S., Chau Y.K., Luton L., Bergiut G.A., Swaine D.J. Methylation of Arsenic in the Aquatic Environment, Conference Proceedings on Trace Substances in Environmental Health-XI, Hemphill University Missouri (1977)
48. Grove J.R. In 'Lead in the Marine Environment' D. Silverfobb, H. Branica, Z. Konrad (Editors) Pergamon Press, Oxford, U.K. pp 45–52 (1980)
49. Muddock B.G., Taylor D. 'The acute toxicity and bioaccumulation of some alkyl lead compounds in marine animals'. In *International Experts Discussion Meeting Lead-occurrence, Fate and Pollution in the Marine Environment*. Rovins, Yugoslavia (1977).
50. Thayer J.S. *Occurrence—Biological Methylation of Elements in the Environment*. American Chemical Society, Washington DC. ACS Advances in Chemistry Series No 182 188 (1978)
51. Grandjean P., Nielson T. *Residue Review* **72** 97 (1979)
52. Wong P.T.S., Silverberg B.A., Chau Y.K., Hodson P.V. *'Lead and the Aquatic Biota'* In *Biogeochemistry of Lead*, editor J. Nriagu, Elsever Press, New York Chapter 17, pp 279–342 (1978)
53. Potter H.R., Jarview A., Markell R.N. *Water Pollution Control* **76** 123 (1977)
54. Wong P.T.S., Chau J.K., Kramer O., Bengett G.A. *Water Research* **15** 621 (1981)
55. Siroto C.R., Uthe J.E. *Analytical Chemistry* **49** 823 (1977)
56. Chau Y.K., Wong P.T.S. 'Lead in the marine environments' M. Branica, M. Konrad editors, In *Proceedings of the International Expert Discussions on Lead Occurrence Fate and Pollution in the Marine Environment* Rovini, Yugoslavia (1977), Pergamon Press, New York (1980)
57. Reamer D.O., Zoller W.M., O'Haver T.C. *Analytical Chemistry* **50** 1449 (1978)
58. Moi E.D., Beccaria A.M., A Dehydration method to avoid loss of trace elements in biological samples In *Proceedings of the International Experts Discussion on Lead—Occurrence, Fate and Pollution in the Marine Environment* Rovini, Yugoslavia (1977)
59. W.H.O. Task Group. Sharrat M. (Chairman) Vouk V.B. (Secretary) *Environmental Health Criter* **51** 1 (1980)
60. Zuckerman J.J., Residorf P.R., Ellis H.V., Wilkinson R.R. In *Organometals and Organometalloids, Occurrence and Fate in the Environment*. Brinkman F.E., Bellama J.M. Editors, *ACS Symposium. Series No. 28*, American Chemical Society Washington DC pp 388–422 (1978)
61. Bock R. Residue Review **79** 216 (1981)
62. Gächter R., Muller H. *Handbuch der Kunstoff—Additive Hanser*, Munich (1979)
63. Fishbein L. *Science of the Total Environment* **2** 341 (1974)

64. Meinema H.A., Burger-Wiersma T., Versluis-de Haan S., Gevers E.C. *Environmental Science and Technology* **12** 288 (1978)
65. Laughlin R.B., French W., Johannson O., Brinkman F.E. *Chemosphere* **13** 575 (1984)
66. Maguire D.J., Carey J.H., Hale E.J. *J. Agriculture and Food Chemistry* **31** 1060 (1983)
67. Getzendanner, M.F. Corbin, H.B. *J. Agriculture and Food Chemistry* **20** 881 (1972)
68. Guard H.E., Cabet A.B., Coleman W.M. *Science* **213** 770 (1981)
69. Jackson J.E., Brinkman F.E., Iverson W.P. *Environmental Science and Technology* **16** 110 (1982)
70. Rapsonankes S., Weber H. *Environmental Science and Technology* **19** 352 (1985)
71. Van Nguyen V., Posey J.J., Eng G. *Water, Air and Soil Pollution* **23** 417 (1984)
72. Byrd J.T., Andreae M.O. *Science* **218** 565 (1982)
73. Braman R.S., Tompkins M.A. *Analytical Chemistry* **51** 12 (1979)
74. Hodge V.F., Seidel S.L., Goldberg E.D. *Analytical Chemistry* **51** 1256 (1979)
75. Lasko J.M., Peoples S.A. *J. Agricultural and Food Chemistry* **23** 674 (1975)
76. Hallas, L.E, Means, J.C., Cooney, J.J. *Science* **215** 1505 (1982)
77. Tugrul S., Balkas T.I., Goldberg E. *Marine Pollution Bulletin* **14** 297 (1983)
78. Gilmour C.C, Tuttle J.H., Means J.C. In *Marine and Estuarine Geochemistry* Sigleo A.C., Haltori A. Editors, Lewis Publishers, Chelsea pp 239-258 (1985)
79. Gilmour C.C., Tuttle J.H. *Analytical Chemistry* **58** 1848 (1986)
80. Skopintsev B.A. *Oceanography* **6** 361 (1966)
81. Williams P.J. 'Determination of organic components' In *Chemical Oceanography* J.P. Riley, G. Skirrow (Editors), volume 3. Academic Press, New York. pp 443-477 (1975)
82. Skopintsev B.A. *Oceanography* **16** 630 (1976)
83. Wangersky P.J., Zika R.G. *The Analysis of Organic Components in Seawater. Report 3* NRCC 16566, Marine Analytical Chemistry Standards Programme (1978)
84. Van Hall C.E., Stenger V.H. *Water Sewage* **111** 266 (1964)
85. Van Hall C.E., Barth D., Stenger V.A. *Analytical Chemistry* **37** 769 (1965)
86. Van Hall C.E., Safrenko J., Stenger V.A. *Analytical Chemistry* **35** 319 (1963)
87. Gotterman H.L. In *Methods for Chemical Analysis of Freshwater* Blackwell, Oxford pp 133-145 (1969)
88. *Water Pollution Research Laboratory, Stevenage*, Herts, U.K. Notes on Water Pollution No 59. HMSO., London (1972)
89. Salvalla, Battori N., Ribas Soler F., *Oronic Durich J. Doc. Invest Hidrol* **17** 303 (1974)
90. Zischke J.A., Arthur J.W. *Archives of Environmental Contamination and Toxicology* **16** 225 (1987)
91. Guidici M., De N Migliore S.M., Guarino S.M. Gambardella C. *Marine Pollution Bulletin* **18** 454 (1987)
92. Hall L.W., Bushong S.J., Ziegerofuss M.C., Johnson W.E., Herman R.L., Weight D.A. *Water, Air and Soil Pollution* **39** 365 (1988)
93. Minchin D., Duggan C.B., King W. *Marine Pollution Bulletin* **18** 604 (1987)
94. Langston W.J. Burt G.R., Mingjiang Z. *Marine Pollution Bulletin* **18** 634 (1987)
95. Pickwell G.V. Steinert S.A. *Marine Environmental Research* **24** 215 (1988)
96. Roberts M.H. *Bulletin of Environmental Contamination and Toxicology* **39** 1012 (1987)
97. Weis J.S., Gottlieb J., Kwintkowski J. *Archives of Environmental Contamination and Toxicology* **16** 321 (1987)
98. His E., Robert R. *Marine Biology* **95** 83 (1987)
99. Bailey S.K., Davies I.M. *Environmental Pollution* **55** 161 (1988)

100. Goodman L.R., Cripe G.M., Moody P.H., Halsell D.G. *Bulletin of Environmental Contamination and Toxicology* **41** 746 (1988)
101. Webb J.L. In 'Enzyme and Metabolic Inhibitors' Vol. 3, chapter 6, Academic Press, New York (1966)
102. Stauffer R.E., Bull J.W., Jenne E.A. *Chemical Studies of Selected Trace Elements in Hot Spring Drainages of Yellowstone National Park*, Geological Survey Professional Paper 1044F. US Government Printing Office, Washington DC (1980)
103. Andreae M.O. *Analytical Chemistry* **48** 820 (1977)
104. McGuire R.J., Tracz R.J. *Water Pollution Research Journal of Canada* **22** 227 (1987)
105. Braman R.S. In *'Arsenical Pesticides'* Woolson E.A. Editor, American Chemical Society, Washington DC Series **7** 108 (1975)
106. Cox D.P. In *'Arsenical Pesticides'* Woolso American (Editors), Chemical Society, Washington DC Series **7** 81 (1975)
107. Chapman A.C. *Analyst (London)* **51** 548 (1926)
108. Cockell J.A., Hilton J.W. *Aquatic Toxicology* **12** 73 (1988)

# CHAPTER 4
# TOXICITY TESTING

## 4.1 TEST DETAILS

Toxicity testing on animals, as opposed to humans, is centred on exposing a known number of animals to known concentrations of the suspect toxic substance in their diet and ascertaining the number of mortalities and/or the number of adverse effects (e.g. reduced growth) occurring in a specified time span. The experiments are carried out at a number of different concentrations of the suspect toxicant using a fixed time span of $T$ days in all experiments.

From the data obtained it is possible to derive statistically the concentration of the test substance that will, in $T$ days exposure, kill 50% of the test animals, the so called $LC_{50}$. Also, it is possible to derived the concentration of test substance that will cause a specific adverse effect on 50% of the test animals, the so called $LE_{50}$.

The whole set of experiments can then be repeated for a different test duration $T^1$ days to establish the effect of exposure time on $LC_{50}$ or $LE_{50}$. Comparison of $LC_{50}$ or $LE_{50}$ values for a particular animal for different test substances enables the relative toxicities of different substances to be evaluated whilst comparison of $LC_{50}$ or $LE_{50}$ values for different test animals using the same test substance enable the relative effect of a given test substance on different species to be evaluated.

The test substance can be fed to the animal in a liquid form or as a solid diet, or in the case of fish and crustacea it can be added in controlled amounts to the water in which the tests are conducted. By using seawater and river water it is possible to compare relative effects of particular concentrations of test substances in saline and non saline media.

Additional tests that can be carried out are the measurement of the concentration of test substances in the tissues and organs of the test animal in which many test substances concentrate to a much higher concentration than is present in the surrounding water (see Chapter 5—bioaccumulation). Correlations can then be obtained between $LC_{50}$, $LE_{50}$ and actual concentrations in animal tissues and organs.

Such tests are, of course, only applicable to sacrificial animals and not to humans. In the latter case, all that can be done is to test the suspect toxicant on a range of laboratory animals (e.g. mice, monkeys) that it is perceived will react to the test substance similarly to man and to apply to the $LC_{50}$ and $LE_{50}$ values so obtained a safety factor (usually 1/10 to 1/100) such that the toxic effect, if any, to man will be absent or minimal. This practice is used extensively in drug evaluation.

## 4.2 EFFECT OF TEST SUBSTANCE CONCENTRATION ON $LC_{50}$ AND $LE_{50}$. DETERMINATIONS ON FISH AND CRUSTACEA

Short term $LC_{50}$ or $LE_{50}$ tests are run for 4 days, i.e. 4 days $LC_{50}$ or $LE_{50}$. Commonly, to obtain more complete information tests might also be run for 1, 10, 100 and 1000 days. In Figure 4.1 is illustrated the effect of test substance concentration on $LC_{50}$ for tests run for 4 days and 10 days i.e. 4 days $LC_{50}$ and 10 days $LC_{50}$. From this plot it is seen that about 390 µg $L^{-1}$ of the test substance would cause 50% mortality of fish in 4 days reducing to 115 µg $L^{-1}$ during a 10 days exposure.

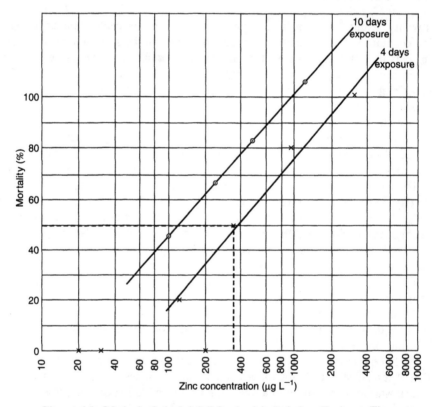

**Figure 4.1.** Method of obtaining $LC_{50}$ by interpolation. Toxicant: Zinc.

## 4.3 EFFECT OF EXPOSURE TIME ON $LC_{50}$ AND $LE_{50}$ DETERMINATIONS ON FISH AND CRUSTACEA

In Figure 4.2 is shown the effect of exposure time in days up to 1000 days in the $LC_{50}$ test on % mortalities of salmon and non-salmonid fish. As shown in Table 4.1

# TOXICITY TESTING

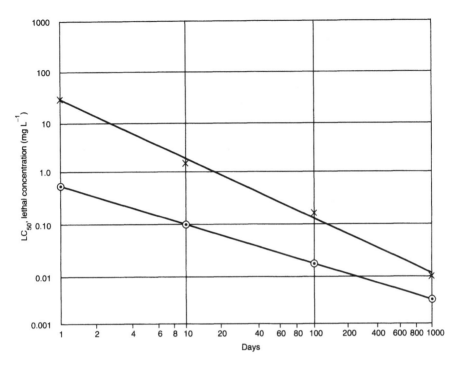

**Figure 4.2.** Effect of exposure time of fish to cadmium on $LC_{50}$; ×: non salmonid fish; O: salmonid fish.

**Table 4.1.** Effect of exposure time of fish on $LC_{50}$.

| Exposure time (days) in $LC_{50}$ test | Approx $LC_{50}$ (µg L$^{-1}$) | |
| --- | --- | --- |
| | Non salmonids | Salmonids |
| 1 | 30 | 0.5 |
| 10 | 2 | 0.1 |
| 100 | 0.2 | 0.015 |
| 365 | 0.04 | 0.005 |
| 1000 | 0.01 | 0.003 |

i. Salmonid fish are more sensitive to the test substance than are non-salmonids.
ii. The lower concentration of test substance in the water the longer the fish survive as would be expected.

In the examples illustrated in Figure 4.2 30 µg L$^{-1}$ of test substance will kill 50% of non-salmonid fish in 1 day exposure whilst in water containing 0.04 µg L$^{-1}$ of test substance 50% of the non-salmonid fish will survive a one year exposure.

It is seen above that the toxic effect of a substance on fish is a consequence of not only the concentration of the substance but also the duration of exposure, the lethal ($LC_{50}$) or adverse effect ($LE_{50}$) concentration becoming progressively lower as the duration of exposure increases.

## 4.4 EFFECT OF OTHER EXPERIMENTAL PARAMETERS ON $LC_{50}$ AND $LE_{50}$ DETERMINATIONS

$LC_{50}$ or $LE_{50}$ values reported in the literature for a particular toxicant and particular exposure time, e.g. 4 days $LC_{50}$, vary considerably. An example of the ranges of values obtained for a particular toxicant are given below:

| Duration of Toxicity test, days | Range of $LC_{50}$ values reported in literature |
| --- | --- |
| 1 | 6.6–124 |
| 10 | 1.2–23 |
| 100 | 0.23–4.3 |
| 1000 | 0.05–0.8 |

Such variability in reported $LC_{50}$ values is a consequence of lack of control of experimental parameters when $LC_{50}$ measurements are being made, also the fact that some types of fish are more sensitive to a particular test substance than others. Experimental parameters that effect the results obtained in $LC_{50}$ or $LE_{50}$ measurements include, temperature (Figure 4.3), pH (Figure 4.4), water hardness (Figure 4.5), salinity, dissolved oxygen content, presence or absence of light, water flow rate through test chamber, chemical form of toxicant and acclimatisation of fish to test conditions.

In carrying out $LC_{50}$ or $LE_{50}$ measurements, as many as practicable of these parameters, should be controlled in order to make intermeasurement reproducibility as high as possible and all test parameters should be quoted with the experimental result.

## 4.5 DERIVATION OF MAXIMUM SAFE CONCENTRATION STANDARD (S) OF A TEST SUBSTANCE FOR CONTINUOUS EXPOSURE TO FISH

A plot of the range available $LC_{50}$ data versus exposure time data for a particular test substance and type of fish is shown in Figure 4.6. In selecting potential values of the standard a boundary line (dotted) drawn to enclose the lower limits of the reported lethal effect ($LC_{50}$) or adverse effect ($LE_{50}$) concentration (i.e. a conservative estimate) would describe a continuous standard in the form of an equation predicting the maximum acceptable concentration of the test substance (with no safety margin) permissable for a specified duration of time. Thus, for continuous exposure of fish the 365 days asymptote $S_x = 0.22$ µg L$^{-1}$ (220 µg L$^{-1}$) would

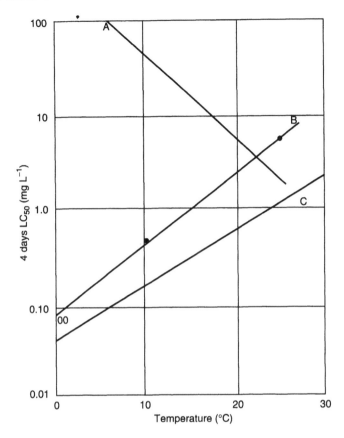

**Figure 4.3.** Effect of test temperature on 4 days $LC_{50}$. A. Silver (as silver nitrate) with salmonid fish, i.e. increase in temperature increases in toxicity. B. Copper with salmonid fish. C. Cadmium with non-salmonid fish. B and C, increase in temperature reduces toxicity. Chromium, zinc, silver with non-salmonid fish: toxicity independent of temperature. Mercury, cadmium, copper and zinc with freshwater invertebrates: increase in temperature increases toxicity.

represent a potential standard ($S$) for the survival of fish for one year in the test water. However, as in practice the concentration of the test substance in the water is not measured continuously it is desirable to apply a safety factor to the $S_x = 220$ µg $L^{-1}$ figure to ensure that any hour to hour variations in concentration of the test substance in the test water does not exceed 220 µg $L^{-1}$. If a safety factor of 10 is applied then this standard should be safe for the longer term (one year) survival of fish unless erratic variations in the concentration of the test substance occur from time to time in the test water. This long term standard ($S_x/10 = 22$ µg $L^{-1}$) incorporating a safety factor of ten might be stated as the annual long term average concentration of test substance in test water. However, adoption of this standard might allow higher concentrations of the test substance in test water to occur for shorter periods during the year and there is a

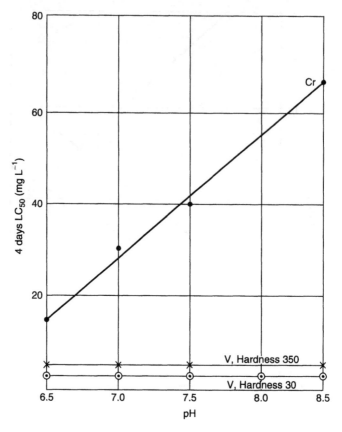

**Figure 4.4.** Effect of pH on $LC_{50}$. Freshwater fish and invertebrates. Toxicant chromium.

potential risk that these excursions in concentration of the test substance would be sufficiently great so as to cause damage to the fish.

To overcome this the 95% percentile concept has been adopted, i.e. that concentration of test substance that could be safely exceeded for 5% of one year (i.e. 17 days per year it daily analysis of the water are performed or 2 weeks per year if weekly analysis of the water are performed).

A plot of the percentage of test substance during which the concentration of test substance in the test water lies between particular concentration limits versus measured concentration is shown in Figure 4.7. Thus, 19% of 365 daily samples, i.e. 69 out of 365 samples have a test substance concentration between 200 and 300 µg L$^{-1}$. 2% of 365 daily samples, i.e. 7 out of 365 samples have a test substance concentration, between 800 and 900 µg L$^{-1}$. Reference to this graph shows that for 5% of the time, i.e. 17 samples out of 365, the concentration of test substance is about 1000 µg L$^{-1}$ i.e. $S_{95} = 1000$ µg L$^{-1}$. Reference to the 17 days exposure results in Figure 4.6 gives an $S_{95}$ value of 0.9 mg L$^{-1}$ (900 µg L$^{-1}$) excluding the safety factor of 10 or $S_{95} = 90$ µg L$^{-1}$ incorporating

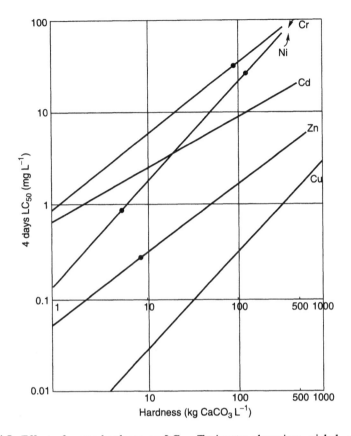

**Figure 4.5.** Effect of water hardness on $LC_{50}$. Toxicants: chromium, nickel, cadmium, zinc and copper.

the safety factor of 10. Thus, providing the concentration of test substance in test water does not exceed $S_{95} = 90$ μg $L^{-1}$ in this example for more than 5% of the exposure time, in this case one year, then all fish will survive unharmed for one year in the test water.

| Daily analysis of test substance in test water number of analysis per annum | | Maximum permissable concentration of test substance in test water for survival of salmonid fish for one year, μg $L^{-1}$ | | | |
|---|---|---|---|---|---|
| | | Without safety factor of 10 | | With safety factor of 10 | |
| | | $S_x$ | $S_{95}$ | $S_x/10$ | $S_{95}/10$ |
| 348 | — | 220 | — | 22 | — |
| — | 17 | — | 900 | — | 90 |

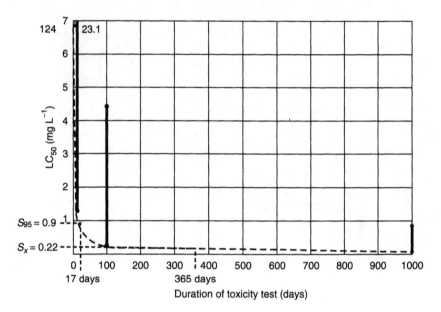

**Figure 4.6.** Test duration–$LC_{50}$ plot for nickel.

The data quoted in Figure 4.6 refers to $LC_{50}$ values, i.e. concentrations of test substances which cause 50% mortality of test animals. The $S_x$ and $S_{95}$ values referred to above refer to maximum permissable concentrations of test substance in test water respectively for 95% and 50% of a year for the survival of fish with nil mortalities. However, we are concerned not only with mortalities in fish but also ill health, i.e. adverse effects. A concentration of test substance in test water at which fish do not die could be too high for them both to survive and remain in good health in order to ensure their continues existance in the ecosystem. Adverse effect concentrations, i.e. $LE_{50}$, $S$ and $S_{95}$, will obviously be lower than the $LE_{50}$, $S$ and $S_{95}$ concentrations quoted above for lethal effect concentrations. Adverse effects include impaired reproduction, reduced rate of growth, abnormal developments and illness.

## 4.6 CUMULATIVE $LC_{50}$ VALUES

Only rarely does a water which is toxic towards fishlife contain a single toxicant. If toxic impurities are present in any appreciable amount then it is likely that several of them will adversely effect fish life. Assuming, as is generally the case, that no synergistic effects exist then the effect of toxicants is additive. The following progressive dilution technique enables the cumulative effect of toxicants on fish to be assessed.

Polluted rivers have been assessed for their toxicity by toxicity tests in flowing water on the river bank using graded dilutions of the river water. Caged fish are

# TOXICITY TESTING

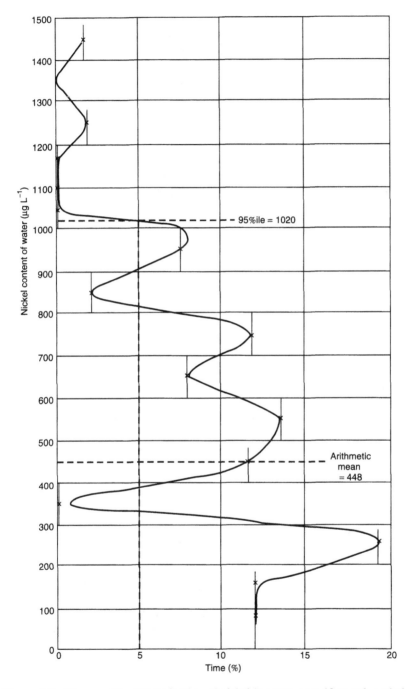

**Figure 4.7.** 95 percentile. Determination of nickel in water over 12 month period.

exposed to the river water for a number of days and the fish mortality rate and pollutant concentrations measured at daily intervals during this period. Simultaneously, caged fish are exposed to a range of dilutions of river water and the same measurements repeated. From the results obtained the dilution causing 50% mortality in 2 days is estimated from various fish species at each location.

Figures 4.8(a)–(d) show test duration–percentage mortality curves obtained from fish in (a) polluted waters and (b) less polluted waters at zero dilution and ×1, ×2, ×5 and ×10 dilutions of river water. From these curves the % of mortality occuring after 2 days exposure for those polluted (a) and less polluted (b) river waters can be obtained. Plots of % mortality versus dilution enables the dilution corresponding to 50% mortality to be read off (Figures 4.9(a) and (b)). From these curves it is seen (Figure 4.9(a)) that for the more polluted water a 50% mortality results when the original river water sample has been diluted ×4 times, and for the relatively unpolluted water B (Figure 4.9(b)) only ×2.8 times dilution is required to achieve the same effect. The results of these studies are presented not as a concentration of pollutants in the rivers, but as cumulative fractions of the relevant laboratory derived 2 days $LC_{50}$ for each species and substance the sum of which is compared with the observed toxicity at each location.

Thus, considering a simple example, if a relatively toxic river water A before dilution contained 50 mg $L^{-1}$ zinc and 10 mg $L^{-1}$ copper then the ×4 dilution of this causing 50% mortality (Figure 4.9(a)) would contain 12.5 mg $L^{-1}$ zinc and 2.5 mg $L^{-1}$ copper, i.e. river derived cumulative 2 days $LC_{50} = 12.5 + 2.5 = 15$ mg $L^{-1}$. Similarly, if a relatively less toxic water B before ×2.8 dilution (Figure 4.9(b)) contained 8 and 1.5 mg $L^{-1}$ of zinc and copper then the fourfold dilution would contain 2.8 and 0.5 mg $L^{-1}$ zinc and copper, i.e. river derived cumulative 2 days $LC_{50} = 2.8 + 0.5 = 3.2$ mg $L^{-1}$. If the laboratory derived 2 days $LC_{50}$ values for zinc and copper are, respectively, 12 and 6 mg $L^{-1}$ i.e. cumulative 2 days $LC_{50} = 18$ mg $L^{-1}$, then the river derived cumulative 2 days $LC_{50}$ as a fraction of the laboratory derived 2 days $LC_{50}$ (i.e. cumulative proportion of laboratory derived 2 days $LC_{50}$) is given by:

$$\text{Polluted Water } A = \frac{2 \text{ days } LC_{50} \text{ river derived}}{2 \text{ days } LC_{50} \text{ laboratory derived}} = \frac{15}{18} = 0.83$$

$$\text{Less Polluted Water } B = \frac{2 \text{ days } LC_{50} \text{ river derived}}{2 \text{ days } LC_{50} \text{ laboratory derived}} = \frac{3.2}{18} = 0.18$$

Observed difference between river derived and laboratory derived cumulative $LC_{50}$ can be ascribed to the effects of factors such as hardness, pH, temperature and dissolved oxygen prior to summation.

This approach has been applied to an assessment of fishery status of rivers where it has been found that if the sum of the proportions of 2 days $LC_{50}$ exceeds about 0.3 then fish will not survive well enough to support fishing activities[2-6]. Table 4.2 shows this effect for a range of river waters of different total hardness.

# TOXICITY TESTING

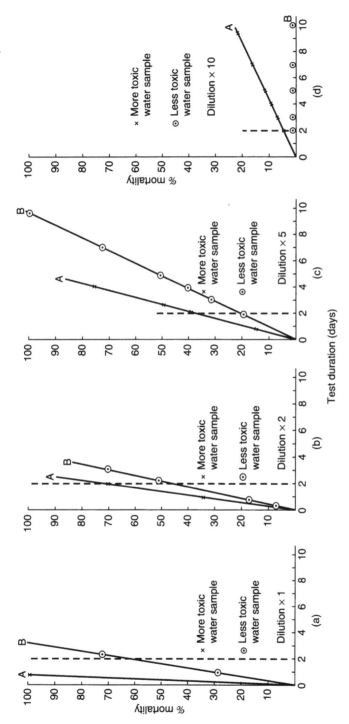

**Figure 4.8.** Test duration–% mortality curves for a range of river water dilutions; ×: a more toxic water, O: a less toxic water.

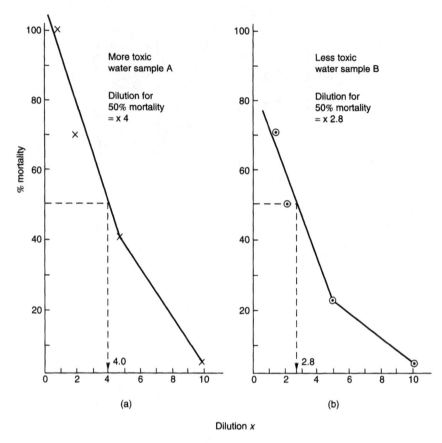

**Figure 4.9.** Dilution–% mortality curves. (a) a more polluted river water; (b) a less polluted river water.

**Table 4.2.** Differences of cumulative proportions of 2 days $LC_{50}$ (laboratory derived) versus fishing status and water hardness.

| Survey | Total Hardness | Median cumulative 2 days $LC_{50}$ (µg L$^{-1}$) | | |
|---|---|---|---|---|
| | | Fishless | Marginal | Fish Present |
| 1 | 11–20 | 0.45 | 0.42 | 0.05–0.25 |
| 2 | 100–170 | 0.1–0.2 | 0.16 | 0.13–0.16 |
| 3 | 100–300 | >0.28 | — | <0.28 |
| 4 | 134–292 | >0.1 | — | <0.1 |
| 5 | 500 | >0.32 | 0.25–0.32 | <0.25 |
| 6 | 70–745 | 0.32–2.95 | 0.3–0.37 | 0.005–0.02 |

# REFERENCES

1. Brown V.M., Shurban D.G., Shaw D. *Water Research* **4** 363 (1970)
2. Alabaster J.S., Garland J.H.N., Hart I.C., Solhe J.F.S. *An Approach to the Problem of Pollution and Fisheries.* Symposium of the Zoological Society London **29** 87 (1972)
3. Hart I.C. *The Toxicity to Fish of Some Rivers in the Yorkshire Ouse Basin* W.P.R. Report 1299 HMSO Water Pollution Research Laboratory (1974)
4. Howells E.S., Howells M.E., Alabaster J.S. *J. Fisheries and Biology* **22** 447 (1983)
5. Brown V.M. *Water Research* **2** 723 (1968)
6. Department of the Environment, Water and the Environment. The Implementation of Directive 76 (464) EEC on Pollution Caused by Certain Dangerous Substances Discharged in to the Aquatic Environment of the Community: Circular 18/85 September, 1985 London, HMSO (1985)

# CHAPTER 5
# Bioaccumulation Processes

## 5.1 INTRODUCTION

It has been observed in Chapter 2 that sediments in rivers and the oceans have the property of adsorbing some types of dissolved substances in the overlying water so that the concentration in the sediment (in mg kg$^{-1}$) is appreciably greater than that in the water (in µg L$^{-1}$) with which the solid is in contact.

A convenient method of expressing this phenomena is by calculating a concentration factor expressed by:

$$\frac{\text{Concentration of substance in sediment (µg kg}^{-1})}{\text{Concentration of substance in water (µg L}^{-1})}$$

Observed concentration factors for a range of organometallic compounds and metal ions in different types of water are tabulated in Table 5.1. Where the concentration factor is appreciably greater than unity the dissolved phase shows a tendency to be adsorbed by the sediment. Examination of the data in Table 5.1 allows the following conclusions to be drawn, see also Figure 5.1:

(a) All the inorganic metal ions listed are strongly adsorbed on to sediments
(b) Organotin compounds are strongly adsorbed on to river water sediments.

The data suggests that organotin compounds may not be as strongly adsorbed on to sediment in saline water i.e. coastal and seawaters as they are in non-saline waters.

(c) Organo compounds of mercury and arsenic are not adsorbed on to sediments in either non-saline or saline waters.
(d) The concentration factors for organolead compounds range from very low values, i.e. no concentration in sediments, to 100, i.e. some adsorption.

Both sources of pollution, i.e. dissolved or sedimentary, are capable of entering living creatures with possible adverse effects. The concentration of toxicants present in sediments is a measure of its concentration in the water over a period of time and is therefore a measure of the risk to creatures. In the case of bottom feeding creatures there is the additional risk of direct ingestion of contaminated sediments in the gills and mouth with consequent adverse effects.

**Table 5.1.** Concentration factors for organometallics and inorganic ions between sediments and liquid phases in water.

| | in Water ($\mu g\ L^{-1}$) | in Sediment ($\mu g\ kg^{-1}$) | | Factor = $\dfrac{\text{Sediment}\ (\mu g\ kg^{-1})}{\text{Water}\ (\mu g\ L^{-1})}$ | | |
|---|---|---|---|---|---|---|
| | | | | Minimum | Maximum | Mean |
| Alkylmercury | 0.06 | <0.01 | Coastal | — | — | <0.17 |
| | 0.006–1.15 (mean 0.57) | <0.01 | Rivers | <0.0086 | <1.66 | <0.02 |
| Inorganic Mercury | 0.009–13.0 (mean 6.5) | 910–46 800 (Mean 23 850) | River | 3600 | 101 110 | 3669 |
| Alkyl lead | <0.00001 | <0.01 | Coastal | — | — | 1000 |
| | 50–530 (mean 290) | <0.01 | Rivers | 0.000019 | 0.0002 | <0.00003 |
| Inorganic lead | 0.02–200 (mean 100) | 23 000–38 200 (mean 30 600) | Coastal | 191 | $1.15 \times 10^6$ | 306 |
| | 0.13–60 (mean 30) | 110–506 000 (mean 253 000) | River | 846 | 8430 | 8433 |
| Monobutyltin | 0.035–0.050 (mean 0.042) | 280 | River | 5600 | 8000 | 6660 |
| | <0.0001–0.3 (mean 0.15) | 1–8 (mean 4.5) | Sea | 26.6 | >10 000 | 30 |
| Dibutyltin | 0.010–0.040 (mean 0.025) | 140 | River | 3500 | 14 000 | 5600 |
| | <0.001–1.6 (mean 0.8) | 0.1–1.0 (mean 0.55) | Sea | 0.62 | >100 | 0.69 |
| Tributyltin | 0.005–0.015 (mean 0.010) | 55 | River | 3667 | 11 000 | 5500 |
| | 0.06–0.78 (mean 0.42) | 0.01–0.3 (mean 0.15) | Sea | 0.17 | 0.38 | 0.36 |
| Inorganic tin | <0.0001 | 1000–20 000 (mean 10 500) | Coastal | >$10^7$ | >$2 \times 10^8$ | >$1.05 \times 10^8$ |
| Alkylarsenic | 2.5–2.6 (mean 2.55) | <0.01 | Coastal | <0.0038 | <0.004 | <0.0039 |
| Inorganic arsenic | 1.00–1.04 (mean 1.02) | 1600–117 000 (mean 59 300) | Coastal | 1600 | 112 500 | 58 140 |
| | 0.42–490 (mean 245) | 220–28 000 (mean 14 100) | River | 57.1 | 523 | 57.6 |
| Inorganic copper | 0.069–9.7 (mean 4.85) | 5400–84 800 (mean 45 100) | Coastal | 8742 | 78 260 | 9298 |
| | 0.11–200 (mean 100) | 70–244 000 (mean 122 000) | River | 636 | 1220 | 1220 |
| Inorganic nickel | 0.2–15.0 (mean 7.6) | 30 000–57 000 (mean 43 500) | Coastal | 3800 | $1.5 \times 10^5$ | 5723 |
| | 1.5–4.5 (mean 3.0) | 1000–238 000 (mean 119 500) | River | 666 | 52 888 | 39 883 |
| Inorganic selenium | <0.01–0.08 (mean 0.04) | 1500–9 000 (mean 5250) | Coastal | 112 500 | 150 000 | 131 250 |

# BIOACCUMULATION PROCESSES

**Table 5.1.** (*continued*)

|  | in Water ($\mu g\ L^{-1}$) | in Sediment ($\mu g\ kg^{-1}$) |  | Factor = $\dfrac{\text{Sediment } (\mu g\ kg^{-1})}{\text{Water } (\mu g\ L^{-1})}$ | | |
|---|---|---|---|---|---|---|
|  |  |  |  | Minimum | Maximum | Mean |
| Inorganic antimony | 0.30–0.82 (mean 0.56) | 6200–134 000 (mean 9 800) | Coastal | 20 666 | 163 414 | 17 500 |
|  | 0.08–0.42 (mean 0.25) | 10–2900 (mean 1455) | River | 125 | 6904 | 5820 |
| Inorganic manganese | 0.35–250 (mean 125) | 21 800–750 000 (mean 386 000) | Coastal | 3000 | 62 285 | 3088 |

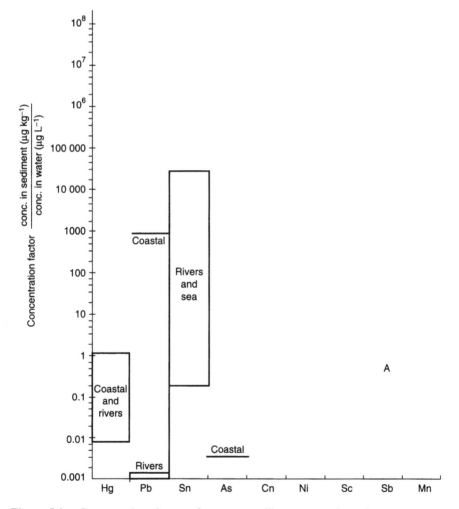

**Figure 5.1a.** Concentration factor of organometallic compounds and corresponding inorganic metal ions in various matrices. A: Organometallic compounds. B: Inorganic metal ions.

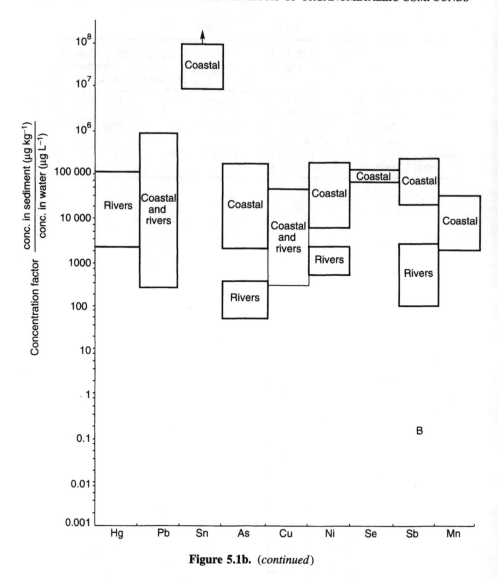

**Figure 5.1b.** (*continued*)

When a creature is exposed to toxicants in the water or sediments in which it lives then the concentration of those toxicants in its tissues gradually increases as a function of exposure time and the concentration of toxicant in the water until the concentration in the tissues of the creature is many times that present in the water. This phenomenon is known as bioaccumulation, which is not to be confused with bioamplification, a process whereby an increase in toxicant levels occurs along a food chain, e.g. plants—minute creatures—fish, as occurs for example in the case of chlorinated insecticides.

In bioaccumulation it is found that the concentration of toxicant in the tissues and particularly in some of its organs such as kidney, liver and opercle, increases with both concentration in the water and exposure time. Measurement of toxicant levels in tissues or organs provide an indication of the amount of exposure to toxicants that the creature has suffered over a period of time. Only limited information is yet available relating concentrations of toxicant in tissues and the onset of ill health or mortality. This is clearly an area where much further work remains to be done.

Monitoring of bioaccumulation of fresh and tidal waters as trends in spacial monitoring as two purposes:

(a) Macroscale, i.e. the identification of potentially unknown areas of elevated contamination and assessment of the extents of the zone of contamination.
(b) Monitoring of bioaccumulation in fresh and tidal waters as trends in time. These need to be monitored to identify trends in contamination, especially near effluent discharges, in order to identify stability, improvement or deterioration in contamination levels.

Spatial and time monitoring programmes of the type discussed above will also give information needed to assess the risk to top predators including man in a particular ecosystem.

A particularly useful way of expressing bioaccumulation is to calculate a concentration fact $C$ which takes into account the concentrations of toxicants in water and the creatures;

$$C = \frac{\text{concentration in creatures } (\mu g \text{ kg}^{-1})}{\text{concentration in water } (\mu g \text{ L}^{-1})}$$

## 5.2 BIOACCUMULATION OF INORGANIC METAL IONS

The results quoted in Table 5.2 demonstrate the increase in toxicant levels found in Rudd fish organs compared to the levels found in the surrounding waters. After 10 weeks exposure to water containing 11 $\mu g \text{ L}^{-1}$ inorganic copper the opercle, liver and kidney contain, respectively, 12 000, 7000 and 6000 $\mu g \text{ kg}^{-1}$ of copper, i.e. concentration factors of 1090, 636 and 545 in these three organs. Thus, analysis of the organs provides a very sensitive method of ascertaining the cause of death when fish kills occur in the environment.

Examination of Rudd fish muscle after exposure to copper is not nearly as sensitive a method of indicating copper pick-up by the fish. Thus, when fish are exposed for 12 hours to water containing 250 $\mu g \text{ L}^{-1}$ copper the concentration of copper found in the muscle is 2000 $\mu g \text{ kg}^{-1}$, i.e. a concentration factor of only 8, whereas the concentrations found, respectively, in opercle, liver and kidney were 52 000, 22 000 and 30 000 $\mu g \text{ kg}^{-1}$, i.e. concentration factors of 208, 88 and 120.

**Table 5.2.** Effect of copper content of rudd fish organs on mortality.

| Exposure Time | 10 weeks | 3 weeks | <12 h | >12 h | Maximum concentration found in organs of wide variety of creatures | | Minimum concentration found in organs of wide variety of creatures | |
|---|---|---|---|---|---|---|---|---|
| | | | | | ($\mu g\ kg^{-1}$ dry weight) | Comments | ($\mu g\ kg^{-1}$ dry weight) | Comment |
| Concentration of metal in water $\mu g\ L^{-1}$ (a) | 11 | 50 | 250 | 1600 | — | | — | |
| Concentrations of metals in fish organs $\mu g\ kg^{-1}$ (b) dry weight | | | | | | | | |
| Opercle | 12 000 | 31 000 | 52 000 | 104 000 | 124 000 | mortalities during 12h exposure | — | no mortalities |
| Liver | 7000 | 20 000 | 22 000 | 40 000 | 48 000–62 000 | | 2000 | |
| Kidney | 6000 | 28 000 | 30 000 | 100 000 | 6000 | | 700 | |
| Muscle | — | — | 2000 | 4000 | — | | — | |
| Concentration factors | | | | | | | | |
| $\dfrac{\text{Concentration organ}}{\text{concentration water}} = \dfrac{b}{a}$ | | | | | | | | |
| Opercle | 1090 | 620 | 208 | 65 | — | — | — | — |
| Liver | 636 | 400 | 88 | 25 | — | — | — | — |
| Kidney | 545 | 560 | 120 | 62.5 | — | — | — | — |
| Muscle | — | — | 8 | 2.5 | — | — | — | — |
| Condition of animal at end of test | good | good | 100% mortality | 100% mortality | — | — | — | — |

Exposure of Rudd to 250 µg kg$^{-1}$ copper in the water for 12 hours causes 100% mortality. The sensitivity of using the opercle to diagnose copper contamination (52 000 µg kg$^{-1}$ copper in organ, concentration factor 208) as opposed to using fish muscle (2000 µg kg$^{-1}$ copper, concentration factor 8) is very apparent. Fish gills and skin, similarly, are poor indicators of copper contamination, i.e. low concentration factors.

The effect of copper content of the water is illustrated in Table 5.3 by data for fish exposed for 12 h to, respectively water containing 250 µg L$^{-1}$ and 1600 µg L$^{-1}$ copper. The copper contents of the opercle are, respectively, 52 000 and 104 000 µg kg$^{-1}$ i.e. concentration factors of 208 and 65.

The copper contents of the muscle when exposed respectively to 250 and 1600 µg L$^{-1}$ copper in water were 2000 and 4000 µg kg$^{-1}$, i.e. concentration factors of 8 and 2.5. concentrations of copper of 250 and 1600 µg L$^{-1}$ in water for 12 hours exposure cause 100% mortalities in fish.

The results in Table 5.3 show the combined effect of exposure time to the contaminated water and the copper content of the water on the occurrence of bioaccumulation. This illustrates that an increase in both of these parameters increases the copper content of the organ.

Experiments with brook trout have been conducted in which the trout were exposed to waters containing 0.09, 0.29 and 0.93 µg L$^{-1}$ inorganic mercury for 9 months and the gonads of the fish then analysed for mercury. Concentration factors so obtained are quoted in Table 5.4. Mortalities can occur of bivalve molluscs, crustacea and fish when they are exposed to 0.1 to 10 µg L$^{-1}$ of inorganic mercury even during short term (4–14 days) exposure (see Table 3.1). Some fatalities, or at least signs of ill health, would be expected to occur under the conditions quoted in Table 5.4, i.e. where mercury contents of gonads have increased to 900 to 12 300 µg kg$^{-1}$ and concentration factors are between 10 000 and 13 000. Comparison of these concentration factors attained during 9 months continuous exposure to mercury with the values of 545 to 1090 attained in the case

Table 5.3. Effect of exposure time and copper content of water on copper content of fish organs.

| | Copper content of opercle (µg kg$^{-1}$) | | | | Copper content of muscle (µg kg$^{-1}$) | | | |
|---|---|---|---|---|---|---|---|---|
| Exposure time (days) | 0.5 | 0.5 | 21 | 70 | 0.5 | 0.5 | 21 | 70 |
| Copper content of water µg L$^{-1}$ (a) | 1600* | 250* | 50** | 11** | 1600* | 250* | 50** | 11** |
| Copper content of organ, µg kg$^{-1}$ (b) | 104 000 | 52 000 | 31 000 | 12 000 | 4000 | 2000 | — | — |
| Concentration factor (b/a) | 65 | 208 | 620 | 1090 | 2.5 | 8 | — | — |

*100% fish mortalities,
**No fish mortalities

**Table 5.4.** Bioaccumulation of inorganic mercury in brook trout, 9 months exposure.

| Mercury content of water, µg L$^{-1}$ (b) | 0.09 | 0.29 | 0.93 |
|---|---|---|---|
| Mercury content of gonads, µg kg$^{-1}$ (a) | 900 | 2900 | 12 300 |
| Concentration factor (a/b) | 10 000 | 10 000 | 13 226 |

of copper (Table 5.2) during 10 weeks exposure is interesting in that it follows the expected trend of increase of concentration factor with exposure time.

Fish in contact with water containing 0.01 µg L$^{-1}$ inorganic mercury and sediments containing 30 µg kg$^{-1}$ of inorganic mercury have been found to contain 341 µg kg$^{-1}$ of mercury in their flesh, i.e. a 34 000 bioaccumulation factor. At Minamata Bay, Japan, mercury levels in some fish attained 50 000 µg kg$^{-1}$ wet weight whilst levels around 20 000 µg kg$^{-1}$ inorganic mercury were common.

Just as fish organs demonstrate the phenomenon of bioaccumulation so do phytoplankton and, indeed, freshwater algae. This makes these species very sensitive indicators of the occurence of pollution in waters. It is seen in Table 5.5 that mean concentration factors range from 155 (arsenic in algae) to 8615 (mercury in phytoplankton).

The bioaccumulation factors obtained for copper (400 in freshwater phytoplankton and 3550 in freshwater algae) compare with the values of 545 to 1090 obtained for fish organs in long term (10 weeks) exposure to copper (Table 5.2).

**Table 5.5.** Bioaccumulation of inorganic metal ions in freshwater phytoplankton and algae.

|  | Mercury | | Arsenic | | Copper | |
|---|---|---|---|---|---|---|
| Concentration of metal in water µg L$^{-1}$ (a) | 0.009–13.0 | | 0.42–490 | | 0.11–200 | |
| Concentration of metal in freshwater Algae µg kg$^{-1}$ (b) | — | | 20 000–56 100 | | 50 000–660 000 | |
| Concentration of metal in freshwater Phytoplankton µg kg$^{-1}$ (c) | 31 200–81 000 | | — | | 40 000 | |
|  | Range | Mean | Range | Mean | Range | Mean |
| Concentration factor | | | | | | |
| Freshwater algae (b/a) | — | — | 114.5–47 619 | 155 | 3300–454 500 | 3550 |
| Freshwater phytoplankton (c/a) | 6231–3.4 × 10$^6$ | 8630 | — | — | 200–363 630 | 400 |

## 5.3 BIOACCUMULATION OF ORGANOMETALLIC COMPOUNDS

### 5.3.1 ORGANOLEAD COMPOUNDS

Bioaccumulation has been observed in the case of tetramethyl lead in rainbow trout (Table 5.6). Analysing whole fish tissues one day and one week after exposure to water containing tetramethyl lead gave concentration factors of 124 (1 day exposure) increasing to between 500 and 934 (one week exposure), i.e. the tetramethyl lead content of the tissues in µg kg$^{-1}$ were between 124 and 934 times greater than that of the water in which the trout lived.

Table 5.6. Accumulation of tetramethyllead in rainbow trout.

| Exposure day | Wt. of fish, g. | Fish alive or dead | Water averaged, µg/L | Fish µg/kg wet wt | Concn. factors[a] |
|---|---|---|---|---|---|
| 1 | 0.1211 | dead | 3.46 | 430 | 124 |
| 2 | 0.3661 | dead | | 1080 | 312 |
|   | 0.7982 | dead | | 2000 | 578 |
| 3 | 0.4116 | dead | | 1320 | 382 |
|   | 0.6300 | dead | | 2090 | 604 |
| 7 | 1.3045 | alive | | 2940 | 850 |
|   | 1.5466 | alive | | 3230 | 934 |
|   | 0.8100 | alive | | 2250 | 650 |
|   | 0.4926 | alive | | 1730 | 500 |

[a] Concentration factor = Concentration of Me$_4$Pb in fish µg kg$^{-1}$/concentration of Me$_4$Pb in water, µg L$^{-1}$

### 5.3.2 ORGANOMERCURY COMPOUNDS

The Mussel Watch Programme is run by the US Environmental Protection Agency. This is a programme in which caged mussels are immersed in environmental waters. A small number of mussels are removed periodically from the cage at known time intervals from the start of the experiment, homogenised, and analysed for the toxicant of interest. A sample of the surrounding water is also taken so that concentration factors in the mussel (bioaccumulation) can be calculated. The experiment is statistically designed to detect 10% changes in toxicant concentrations in the mussels with a confidence of 90%.

In one such programme, Mylitus mussels were suspended in cages in the Firth of Forth. The methyl mercury concentration of the mussels increased from less than 10 µg kg$^{-1}$ one week after the start of the experiment to 60 to 80 µg kg$^{-1}$ after 21 weeks exposure to water with a mean methyl mercury content of 0.06 µg L$^{-1}$. Calculated concentration factors were <160 (<10/0.06) at one week from the start of the experiment to 1170 to 1333 (70/0.06 to 80/0.06) after 21 weeks exposure. These values are similar to those quoted above for tetramethyl lead, viz. a concentration factor of 124 after one day exposure and 500 to 850 after one week's exposure.

Richman et al.[1] have pointed out that mechanisms responsible for higher concentrations of mercury in fish from acidic lakes were poorly understood. However, several hypotheses have been proposed: mercury might enter the catchment with acid deposition: acidification might mobilise mercury bound in lake sediment and catchment soils; lower pH could favour the production of the more bioavailable monomethyl mercury species; pH could influence the rates of mercury methylation and/or demethylation by microorganisms; biotic characteristics of acid lakes could influence mercury transfer and biomagnification; acidification might directly effect lake biota altering the ability of organisms to bioaccumulate and/or excrete mercury. Evidence for and against these hypotheses are discussed, and it was concluded that mercury cycling and uptake in aquatic systems were governed by a variety of interconnecting and sometimes covarying factors, the relative importance of which could differ from lake to lake.

### 5.3.3 ORGANOTIN COMPOUNDS

Tsuda et al.[2] have discussed results obtained in the determination of bioconcentration factors in carp (*Cyprinus carpio*) and the octanol/water partition coefficients for triphenyltin chloride, diphenyltin dichloride, and monophenyltin trichloride. The further metabolism of triphenyltin chloride in the fish was also investigated and the concentrations of phenyltin found in various tissues are discussed.

Anil and Wagh[3] collated samples of barnacles monthly (March 1983–May 1984) from two stations (the shipyard and the harbour) of the Zuan estuary. Total copper and zinc concentrations in the water were 1–11 and 13–46 $\mu g\ L^{-1}$, respectively. Copper concentrations in barnacles were 47 300–864 800 $\mu g\ kg^{-1}$ in the 0.1 cm size group and 39 700–625 700 $\mu g\ kg^{-1}$ in the 1–2 cm size group. Corresponding zinc concentrations 203 600–1 937 500 and 204 300–384 300 000 $\mu g\ kg^{-1}$. Concentration factors were 7060–384 300 for copper and 10 660–84 600 for zinc. In general, concentrations factors in the 0.1 cm size group were higher than those in the 102 cm size group.

Langston and Zhou[4] collected samples of tellnid clams (*Macoma balthica*) from Whitehaven, Cumbria, in June 1984 and acclimated them for use in laboratory experiments on the bioaccumulation and bioelimination of cadmium by this species. At cadmium concentrations of 100 $\mu g\ L^{-1}$ accumulation by the soft tissues were linear (350 $\mu g\ kg^{-1}\ d^{-1}$, dry weight) throughout a 29 day exposure. Amounts of cadmium accumulated by the shell were low and elimination rates high (retention half life 7 days) compared to soft tissues (retention half life 70 days). Gelchromatographic profiles of cytosol extracts from control and experimental groups of clams provided no evidence for the involvement of either metallothionein or metallothionein-like proteins in cadmium accumulation. Most cadmium was bound to high molecular weight ligands although the small amount (less than 15%) associated with low molecular weight ligands might be important in regulating cadmium uptake and elimination phases. The absence of a recognised detoxifying system in this species might be compensated for by the slow rate of cadmium accumulation.

Lyngby and Brix[5] investigated heavy metal contamination in the Limfjord, Denmark, and eelgrass (*Zostera marina*) and the marine mussels (*Mytilus edulis*) and compared these as indicators of heavy metals in shallow coastal areas. Background levels and threshold values were calculated for the organisms and sediments. Significant elevations of heavy metal concentrations were found in the Nissum Broad (mercury), in Veno Bay (cadmium) and Aalborg (mercury, zinc, lead and copper). Positive correlations between concentrations of mercury, lead, cadmium and zinc were found in eelgrass leaves and root rhizomes, mussels, and sediment, but the copper concentrations in mussels did not correlate.

Higgins and Mackey[6] collected samples of *Ecklonia radiata* bimonthly, during July 1982–March 1984, from the shallow sublittoral zone of the kelp beds in Port Hacking estuary, Australia. Tissue concentrations of iron and manganese were approximately 60% higher in late summer compared with the rest of the year. Zinc, cadmium, copper, potassium, calcium, magnesium and sodium concentrations showed no seasonal variations. Concentration factors ranged from 4.0 (sodium) to 68 000 (iron). Seasonally averaged concentrations of sodium, magnesium, calcium and potassium were relatively uniform throughout the kelp tissues. Concentrations of iron, manganese and zinc were highest in the extremities (eroding tip, holdfast tissue), and lowest in the meristematic tissue. Cadmium concentrations were elevated in the extremities, but uniformily distributed in the other tissues. Copper concentrations were highest in holdfast tissue, and lowest in the eroding tip. Experiments using EDTA indicated that approximately 90% of total cadmium and zinc was associated with the apparent free space, whereas corresponding values for copper and iron were 25 and 7% respectively. In view of the rapid exchange between seawater and the apparent free space, it was concluded that Fcklonia *radiata* would not be of general value in the assessment of long term integrated changes of metals in the water column.

## 5.4 GENERAL CONCLUSIONS

A summary of the available data in Table 5.7 shows that, in general, bioaccumulation of inorganic metals on water creatures such as fish and invertebrates and plant life in the rivers and oceans is higher than in the case of bioaccumulation of organometallic compounds, at least in the case of mercury and lead. As shown in Table 5.1 this is also the case in the adsorption of inorganic metals on to river and coastal sediments, i.e. adsorption factors in the cases of mercury, lead and many other elements adsorbed in to river and ocean sediments are greater in the case of inorganic metals than in the case of corresponding organometallic compounds. In Table 5.8 is a representation of the data in Table 5.7. Table 5.8 lists the bioaccumulation under types of sample, viz. fish, invertebrates and plants. Mean concentration factors are quoted.

**Table 5.7.** Concentration factors of inorganic elements and organometallic compounds in water creatures and plants.

| Element | Type of sample | Inorganic element | | Organometallic compounds | |
|---|---|---|---|---|---|
| | | Range of concentrations factors | Mean concentration factor | Range of concentration factors | Mean concentration factor |
| Lead | Sediment (coastal)* | 191–1.15 × 10$^6$ | 306 | — | 100 |
| | Sediment (river)* | 846–8430 | 8433 | 0.000019–0.0002 | <0.00003 |
| | Rainbow trout | — | — | 124 (1 day exposure) 934 (7 days exposure) (as PbMe$_4$) | — |
| Mercury | Phytoplankton | 6231–3.46 × 10$^6$ | 8630 | — | — |
| | Brook trout | 10 000 (at 0.09 μg L$^{-1}$ Hg in water) 13 276 (at 0.93 μg L$^{-1}$ Hg in water) | — | — | — |
| | Mussels | — | — | 1170–1333 (21 weeks exposure) | <160 (1 day exposure) |
| | Sediment (river)* | 3600–101 100 | 3669 | 0.008–1.66 (as alkyl mercury) | <0.02 |
| | Sediment (coastal)* | — | — | — | <0.17 |
| Arsenic | Algae | 111.4–47 619 | 155 | — | — |
| | Sediment (coastal)* | 1600–112 500 | 58 140 | <0.0038–<0.004 | <0.039 |
| | Sediment (river)* | 57.1–523 | 57.6 | — | — |
| Copper | Phytoplankton | 200–363 630 | 400 | — | — |
| | Algae | 3300–454 500 | 3500 | — | — |
| | Opercle (fish) | 65–620 | 65–208 (0.5 day exposure) 620 (70 day exposure) | — | — |
| | Muscle (fish) | 2.5–8.0 | 5.2 (0.5 day exposure) | — | — |
| | Barnacle | 7060–384 000 | — | — | — |
| | Sediment (coastal)* | 8742–78 260 | 9298 | — | — |
| | Sediment (river)* | 636–1220 | 1220 | — | — |
| Zinc | Barnacle | 10 660–84 600 | — | — | — |
| Cadmium | Clam | — | 101 (29 days exposure) | — | — |
| Iron | Kelp | — | 68 000 | — | — |
| Sodium | Kelp | — | 4 | — | — |

*For comparison

**Table 5.8.** Orders of magnitude in the bioaccumulation of inorganic metals and organometallic compounds by fish, invertebratea and plants.

| Type | Metal | Inorganic metals | | | Organometallic compounds | | |
|---|---|---|---|---|---|---|---|
| | | Factor | Sediments (for comparison) | | Factor | Sediment (for comparison) | |
| | | | River | Coastal | | River | Coastal |
| *Fish* | | | | | | | |
| Trout | Pb | — | 8433 | 306 | 934 | <0.00003 | 100 |
| Trout | Hg | 8630 | 3669 | — | — | <0.02 | <0.17 |
| Fish Muscle | Cu | 5.2 (0.5 day exposure) | 1220 | 9298 | — | — | — |
| Fish Opercle | Cu | 620 (70 day exposure) | | | — | — | — |
| *Invertebrates* | | | | | | | |
| Mussel | Hg | — | 3669 | — | 1251 | <0.02 | <0.17 |
| Barnacle | Cu | 195 530 | 1220 | 9298 | — | — | — |
| Barnacle | Zn | 47 600 | — | — | — | — | — |
| Clam | Cd | 101 (29 day exposure) | — | — | — | — | — |
| *Plants* | | | | | | | |
| Phytoplankton | Hg | 8630 | 3669 | — | — | — | — |
| Algae | As | 155 | 60 | 58 140 | — | — | <0.0039 |
| Algae | Cu | 3500 | 1220 | 9298 | — | — | — |
| Kelp | Fe | 68 000 | — | — | — | — | — |
| Kelp | Na | 4 | — | — | — | — | — |

## 5.4.1 FISH

It has been previously stated in Section 5.2 that a bioaccumulation of inorganic metals by fish from water is greater in the case of fish opercle than in the case of fish muscle. This is supported by the data in Table 5.8 which shows a higher concentration factor of 620 after 70 days exposure of fish opercle to inorganic copper compared with a value of 5.2 obtained for fish muscle. The mean concentration factor obtained in the case of the bioaccumulation of inorganic mercury into whole fish was considerably higher (i.e. mean factor 8630) than was obtained for copper and also about twice the value of 3669 obtained for the adsorption of inorganic mercury by river sediments. The bioaccumulation of inorganic mercury by whole fish (factor 8630) is considerably greater than the bioaccumulation of organomercury (factor <0.2–<0.17).

The bioaccumulation of organolead by trout was relatively low (factor 934) and the adsorption by river and coastal sediments was also very low (factors <0.00003 and 100).

## 5.4.2 INVERTEBRATES

A very high degree of bioaccumulation occurs in the case of inorganic copper (mean factor 195 530) and zinc (mean factor 47 600) in barnacles. Both these values are considerably higher than the concentration factors of 1120–9298 obtained, respectively for the adsorption of inorganic copper by river water and coastal sediments. Bioaccumulation of cadmium by clams is quite low (factor 101 after 2 days exposure).

## 5.4.3 PLANTS

The bioaccumulation of inorganic sodium by kelp was very low (factor 4) whilst that of inorganic iron was very high (factor 68 000). The bioaccumulation of inorganic copper by algae and inorganic mercury by phytoplankton were intermediate (mean factors, respectively, of 3500 and 8630. Bioaccumulation of inorganic arsenic was relatively low (factor 155), a similar order of magnitude to the bioaccumulation of cadmium by clams (factor 101).

The bioaccumulation of inorganic mercury by phytoplankton (mean factor 8630) is considerably greater than the adsorption of inorganic mercury by river sediments (mean factor 3669). The bioaccumulation of inorganic arsenic by algae (mean factor 155) is similar to the adsorbtion of inorganic arsenic by river sediments (mean factor 60), but is considerably lower than the adsorption of inorganic arsenic by sediments in a marine environment (mean factor 58 140). The bioaccumulation of inorganic copper by algae (mean factor 3500) is intermediate between the values obtained for the adsorption of inorganic copper by river and coastal sediments (mean factors respectively of 1220 and 9298).

# REFERENCES

1. Richman L.A., Wren C.D., Stokes P.M. *Water Air and Soil Pollution* **37** 465 (1988)
2. Tsuda T., Nakonishi H., Aoki S., Takebayashi J. *Water Research* **21** 949 (1987)
3. Anil A.C., Wagh A.B. *Marine Pollution Bulletin* **19** 177 (1988)
4. Langston W.J., Zhou M. *Marine Environmental Research* **21** 225 (1987)
5. Lyngby J.E., Brix H. *Science of the Total Environment* **64** 239 (1987)
6. Higgins H.W., MacKay D.J. *Australian Journal of Marine and Freshwater Research* **38** 307 (1987)

# CHAPTER 6
# Analysis of Organometallic Compounds in the Environment

## 6.1 ORGANOMERCURY COMPOUNDS

### 6.1.1 NATURAL AND POTABLE WATERS

**Reduction of organomercury to elemental mercury**

Mercury in water samples can exist in inorganic or organic forms or both. Preliminary degradation of organomercury compounds in the sample to inorganic mercury preparatory to analysis is often necessary because the normal methods of reducing inorganic mercury compounds to elemental mercury with reagents such as stannous chloride do not work with organomercury compounds and hence organomercury compounds are not included in such determinations. Owing to the conversion of $Hg^{2+}$ to $CH_3Hg^+$ in natural water, it is often observed that a high percentage of the mercury is present in the form of organic compounds[1]. Some organic mercurials like $CH_3HgCl$ and $(CH_3)_2 Hg$ may be reduced by a combination of cadmous chloride and stannous chloride, but this method requires large quantities of reductants and the use of strong acid and strong alkali[2].

Organic mercury compounds can be decomposed by heating with strong oxidizing agents such as potassium dichromate or nitric acid-perchloric acid, followed by reduction of the formed divalent mercury to mercury vapour[3].

Both methods are rather time-consuming and not readily suitable for automation. Potassium persulphate has also been used successfully to aid the oxidation of organomercury compounds to inorganic mercury and this forms the basis of an automated method[4].

Sulphuric acid acidified potassium permanganate has also been used to decompose organically bound mercury, prior to reduction with stannous chloride and determination of the evolved mercury by atomic absorption spectroscopy. Bennett et al.[5] showed that acid-permanganate alone did not recover three methylmercuric compounds, while the addition of a potassium persulfate oxidation step increased recoveries to 100%. El-Awady et al.[6] confirmed the low recoveries of methylmercury by acid-permanganate. They showed that only about 30% of methylmercury could be recovered by this method, while the use of potassium persulfate produced complete recovery.

Umezaki and Iwamoto[7] differentiated between organic and inorganic mercury in river water samples. They used the reduction-aeration technique. By using stannous chloride solution in hydrochloric acid only inorganic mercury is reduced whereas stannous chloride in sodium hydroxide medium and in the presence of $Cu^I$ reduces both organic and inorganic mercury. The mercury vapour is measured conventionally at 254 nm. Ions that form insoluble salts or stable complexes with $Hg^4$ interfere.

### Potentiometric titration

Potentiometric titration with standard solutions of dithiooxamide at pH5–6 has been used to estimate less than 100 pg total mercury in water samples. The precision in the range 0.5–1.0 mg $L^{-1}$, mercury is about 4%. The first derivative could be used for end-point determination. A wide variety of ions can be tolerated but silver, copper and chloride interfere, and must be separated in a preliminary step.

### Spectrophotometric methods

Ke and Thibert[9] have described a microdetermination of down to 50 μg $L^{-1}$ of inorganic and organic mercury in river water and seawater. Mercury is determined by use of the iodide-catalysed reaction between $Ce^{IV}$ and $As^{III}$ which is followed spectrophotometrically at 275 nm.

Chau and Saitoh[10] reported a spectrophotometric method for the determination of submicrogram amounts of organic and inorganic mercury in lake water based on dithiozone extraction.

### X-ray fluorescence spectrometry

Braun et al.[11] showed that polyurethane foam loaded with diethylammonium diethyldithiocarbamate is suitable for concentration of trace amounts of organic mercury from potable water samples prior to analysis. Organomercury compounds studied included phenylmercury and methymercury species. The polyurethane discs were then analysed by x-ray fluorescence spectrometry. Preconcentration of mercury prior to the measurement has also been achieved by amalgamation with noble metals[12–14].

Braun et al.[15] used radioisotope induced X-ray fluorescence spectrometry to determine phenylmercury, methylmercury and inorganic mercury in river water after preconcentration on diethylammonium diethyldithiocarbamate-loaded polyurethane foam.

### Neutron activation analysis

Becknell et al.[16] first converted the organomercury chloride using chlorine. The mercury was then concentrated by removal as $HgCl_4^{2-}$ by passing the sample solution (500 mL) adjusted to be 0.1M in hydrochloric acid through a paper filter disc loaded with SB-2 ion exchange resin. The paper, together with a mercury standard, is then heat-sealed in Mylar bags and analysed by neutron activation analysis using the 77 keV x-ray photo-peak from the decay of $^{197}$Hg.

### Nuclear magnetic resonance spectroscopy

Robert and Robenstein[17] carried out indirect detection of Hg$^{199}$ NMR spectra of methyl mercury complexes, e.g. CH$_3$Hg$^{II}$ thiol ligands in environmental waters.

### Flow injection analysis

Hanna and Tyson[18] determined total mercury in waters (and urine) by a flow injection–atomic absorption spectrometric procedure. This procedure involved on-line and off-line oxidation of organomercury species and was capable of determining down to 0.1 ng L$^{-1}$ of these compounds.

### Radio-analytical methods

Radiocaromatographic assay has been used[19] as the basis of a method for determining inorganic mercury and methylmercury in river water.

Stary and Prasilova[20-23] have described a selective radiochemical determination of phenylmercury and methylmercury. These analytical methods are based on the isotope exchange reactions with the excess of inorganic mercury-203 or on the exchange reactions between phenyl-mercury and methyl-mercury chloride in the organic phase and sodium iodide-131 in the aqueous phase. The sensitivity of the methods (0.5 µg L$^{-1}$ in 5 mL sample) is, however, insufficient to determine organomercurials in natural waters.

Stary[24] developed a preconcentration-radioanalytical method for determining down to 0.01 µg L$^{-1}$ of methyl and phenyl mercury and inorganic mercury using 100–500 mL samples of potable or river water. Extraction chromatography and dithizone extraction were the most promising methods for the concentration of organo-mercurials in the concentration range 0.01 to 0.1 µg L$^{-1}$. The dithizone extraction method was used for the preconcentration of inorganic mercury.

### Atomic absorption spectrometry

Various workers have made contributions to the determination of organically bound mercury compounds and inorganic mercury in potable and natural waters by atomic absorption spectrometry.[7,10,25-58]

Baltisbergen and Knudson[26] described a method for the determination of the individual concentrations of mercury II and RHg II in aqueous river waters. The mercury metal produced by reduction with tin II is measured by flameless atomic absorption. Selective reduction of mercury II is achieved in the presence of RHg II in sulfuric acid media using tin II. The total mercury content is then measured after oxidation with acidic hydrogen peroxide solution just before injection into the tin II solution of the mercury analyser. The method is useful in the range 1–15 μg L$^{-1}$ with a standard deviation of 1 μg L$^{-1}$.

Umezaki and Iwamoto[7] differentiated between organic and inorganic mercury in river samples. They used the reduction–aeration technique described by Kimura and Miller[50]. By using stannous chloride solution in hydrochloric acid only inorganic mercury is reduced, whereas stannous chloride in sodium hydroxide medium and in the presence of cupric copper reduces both organic and inorganic mercury. The mercury vapour is measured conventionally at 254 nm. Ions that form insoluble salts or stable complexes with Hg(II) interfere.

Simpson and Nickless[32] have described a rapid dual channel method of cold vapour atomic absorption spectroscopy for determining total mercury. A detection limit of 12.5 mg L$^{-1}$ is claimed.

Lutze[33] has described a flameless atomic absorption method for determining total mercury in waters and Grantham[51] has applied the method to waters.

Goulden and Anthony[34] describe how the sensitivity of an automated cold-vapour atomic absorption method for mercury can be improved by equilibrating the reduced sample with a small volume of air at 90°C, to achieve a detection limit of 1 ng mercury. Inorganic mercury, aryl mercury compounds, and alkyl mercury compounds can be distinguished by changing the chemical reduction system. The authors suggest that their method of chemical speciation can be applied to natural waters.

Graf et al.[52] used sulphuric acid acidified potassium permanganate to decompose organically bound mercury, prior to reduction with stannous chloride and determination of the evolved mercury by atomic absorption spectroscopy.

Kalb[53] used concentrated nitric acid to decompose organomercury compounds in river water samples prior to estimation by flameless atomic absorption spectroscopy. Stannous chloride was used to liberate elementary mercury, which is then vaporised by passing a stream of air (1360 mL min$^{-1}$) through the solution. The air stream passes over silver foil, where mercury is retained by amalgamation and other volatile substances pass out of the system. The foil is heated at 350°C in an induction coil and the air stream carries the mercury vapour through a cell with quartz windows. The atomic absorption at 253.65 nm is measured and the mercury concentration (up to 0.02 mg L$^{-1}$) is determined by reference to a calibration graph.

The Water Research Council UK[54,55] have described a method for the determination of mercury in water in which all forms of mercury are converted to inorganic mercury using prolonged oxidation with potassium permanganate. Elemental mercury is then released using stannous chloride and mercury estimated by cold vapour atomic absorption spectrometry at 253.7 nm. Many of the potential interferences in the atomic absorption procedure are removed by the preliminary digestion–oxidation procedure. The most significant group of interfering substances are volatile organic compounds which absorb radiation in the ultraviolet. Most of these are removed by the pretreatment procedure used and the effect of any that remain are overcome by preaeration. Substances which are reduced to the elemental state by stannous chloride and then form a stable compound with mercury may cause interference (e.g., selenium, gold, palladium and platinum). The effects of various anions, including bromide and iodide were studied. These are not likely to be important interferers. Excellent performance characteristics are presented for this method.

Sampling techniques are described in detail including methods of cleaning sample bottles and fixing the sample with a solution of potassium dichromate in nitric acid. Farey et al.,[36] compared this method with a method in which the sample is brominated to quantitatively convert alkyl and aryl mercury compounds to mercuric bromide. Recoveries of inorganic mercury from distilled water spiked with phenylmercury(II) chloride, thiomersal, ethylmercury(II) chloride, methylmercury(II) chloride, phenylmercury(II) acetate and p-tolymercury(II) chloride were greater than 95% (Table 6.1).

When identical conditions of treatment were used on 50 mL samples of tap water and various river waters and sewage effluents, all with added methylmercury(II) chloride there were similar recoveries after 5 min reaction with bromine reagent (Table 6.2).

Farey et al.[36] claim that their pretreatment compares favourably with an established permanganate–sulfuric acid method. An advantage of the technique is that it can easily be carried out while sampling onsite. The sample is collected in glass bottles containing hydrochloric acid and the bromate–bromide solution is added. A bromination reaction time is then provided from the collection of the sample to the analysis in the laboratory and this is far in excess of that necessary to decompose the organic mercury. In addition as aqueous mercury(II) solutions are stabilised by strong oxidising agents, the oxidising conditions so created will help to preserve the inorganic mercury formed.

Abo-Rady[47] has described a method for the determination of total inorganic plus organic mercury in nanogram quantities in natural water, fish, plants and

**Table 6.1.** Recoveries of inorganic mercury from organomercurials in distilled water following bromination. (From Farey et al.[36] Reproduced by permission of Royal Society of Chemistry.)

| Organomercurial | Organic Hg concentrations in distilled water ($\mu g\ L^{-1}$) | Inorganic Hg found after bromination* ($\mu g\ L^{-1}$) | Recovery (%) |
| --- | --- | --- | --- |
| $C_6H_5HgCl$ | 15.1 | 14.9 | 99 |
| $CH_3C_6H_4HgCl$ | 11.9 | 11.8 | 99 |
| $C_2H_5HgSC_6H_4COONa$ | 9.7 | 9.9 | 102 |
| $C_6H_5HgCOCCH_3$ | 10.0 | 10.0 | 100 |
| $C_2H_5HgCl$ | 15.0 | 14.4 | 96 |
| $CH_3HgCl$ | 14.0 | 14.2 | 101 |

*The figures quoted are the means of results obtained from duplicate experiments. Bromination time 1 min using 1 mL of brominating reagent.

**Table 6.2.** Recoveries of inorganic mercury from $CH_3HgCl$ added to natural waters. (From Farey et al.[36]. Reproduced by permission of Royal Society of Chemistry.)

| Spiked medium (14.8 µg L$^{-1}$ of Hg as $CH_3HgCl$) | 1 mL $KBrO_3$–KBr added | | | | 2 mL of $KBrO_3$–KBr added | | | |
|---|---|---|---|---|---|---|---|---|
| | 1 min reaction | | 5 min reaction | | 1 min reaction | | 5 min reaction | |
| | Inorganic Hg found (µg L$^{-1}$) | Recovery (%) | Inorganic Hg found (µg L$^{-1}$) | Recovery (%) | Inorganic Hg found (µg L$^{-1}$) | Recovery (%) | Inorganic Hg found (µg L$^{-1}$) | Recovery (%) |
| River water | 6.8 | 46 | 13.2 | 89 | 13.8 | 93 | 13.6 | 92 |
| River water | 9.9 | 67 | 14.4 | 97 | 13.9 | 94 | 14.4 | 97 |
| River water | 12.6 | 85 | 14.1 | 95 | 14.1 | 95 | 13.3 | 90 |

sediments. This method is based on decomposition of organic and inorganic mercury compounds with acid permanganate, removal of excess permanganate with hydroxylamine hydrochloride, reduction to elemental mercury with tin and hydrochloric acid, and transfer of the liberated mercury in a stream of air to the spectrometer. Mercury was determined by using a closed, recirculating air stream. Sensitivity and reproducibility of the 'closed system' were better, it is claimed, than those of the 'open system'. The coefficient of variation was 13.7% for water, 1.9% for fish, 4.9% for plant and 5.6% for sediment samples.

Doherty and Dorsett[31] analysed environmental water samples by separating the total organic and inorganic mercury by electro-deposition for 60 to 90 min. on a copper coil in 0.1 M nitric acid medium and then determined it directly by flameless atomic-absorption spectrophotometry. The precision and accuracy are within ±10% for the range 0.1 to 10 $\mu g\ L^{-1}$. The sensitivity is 0.1 $\mu g\ L^{-1}$ (50 mL sample).

Ultraviolet irradiation of the sample prior to analyses is another means of decomposing organomercury compounds to the inorganic form. Goulden and Afghan[56] have used ultraviolet irradiation as a means of sample decomposition. After the photochemical oxidation, the formed inorganic mercury is reduced to metallic mercury in the usual way by stannous chloride. This method reduces the consumption of oxidising agents and thus diminishes considerably the risk of contamination; it also leads to shorter analysis times. Determinations with and without irradiation enable the separate determination of total and inorganic mercury, respectively.

Kiemeneij and Kloosterboer[35] have described an improvement on the Goulden and Afghan photochemical decomposition of organomercury compounds in the $\mu g\ L^{-1}$ range in natural water prior to determination by cold vapour atomic absorption spectrophotometry. The decomposition of the organomercurials is carried out by means of ultraviolet irradiation at a suitable wavelength from small low-pressure lamps containing either Hg, Cd or Zn or a mixture of these metals in their cathodes. These lamps have the strongest lines, respectively at 254, 229 and 214 nm and the cadmium 229 lamp was chosen as the best compromise between high quantum efficiency and avoidance of background absorbance due to naturally occurring organic contaminants in water samples (Figure 6.1).

Kiemeneij and Kloosterboer[35] used toroidal silica irradiation cells which can be placed around the lamp (Figure 6.2). The use of the cell shown in Figure 6.2 obviates the transfer of the sample solution to the aeration bottle of the atomic absorption spectrometric detection system, since a glass fruit is mounted near the bottom of the cell, giving the possibility of aeration through the side tube. The outer walls of the cells were coated with aluminium which in turn was protected by a varnish layer. Determinations with and without irradiation make possible separate determinations of total and inorganic mercury, respectively, in about 20 min. The formed inorganic mercury is determined in the usual way by cold vapour atomic absorption after reduction of divalent mercury to mercury vapour. Comparison of the photochemical with a wet-chemical method

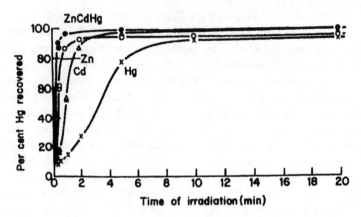

**Figure 6.1.** Percentage recover of mercury from $CH_3HgCl$ added to deionised water as a function of the time of irradiation for various light sources. Concentration 4.1 µg Hg $L^{-1}$; reference $Hg(NO_3)_2$. The small recovery observed for $t = 0$ is probably caused by the thermal or daylight decomposition; it was not observed with diphenylmercury. (From Kiemieneij and Kloosterboer[35]. Reproduced by permission of American Chemican Society.)

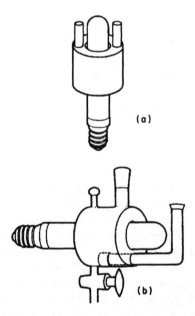

**Figure 6.2.** View of two silica irradiation cells with the light source. Dimensions: height 50 mm, internal diameter 33 mm and external diameter 70 mm. (From Kiemieneij and Kloosterboer[35]. Reproduced by permission of American Chemican Society.)

**Table 6.3.** Comparison of photochemical[35] and wet chemical decomposition[57] of mercury compounds in an acidified natural water sample (River Wall). (From Kiemeneij and Kloosterboer[35] Reproduced by permission of American Chemical Society.)

| Sample treatment | | Hg found, (μg L$^{-1}$) | |
|---|---|---|---|
| | | Ref. 35 | Ref. 57 |
| Unirradiated | | 0.3 | |
| Irradiated | 10 min | 1.01 | 1.00 |
| (ZnCdHg lamp) | 30 min | 1.15 | 1.11 |
| | 30 min | 1.05 | 1.12 |
| Stored with KMn04 | 2% KMn04 | 0.98 (0.06) | 1.02 (0.03) |
| | 4% KMn04 | 1.04 (0.08) | 0.97 |
| | 4% KMn04 | 1.07 | 1.01 |
| Stored with KMn04 | 2% KMn04 | 1.00 (0.24) | 1.06 (0.23) |
| partly evaporated and rediluted | 4% KMn04 | 1.07 (0.27) | 1.06 (0.31) |

(Table 6.3) showed that for organomercury compounds the results of prolonged irradiation compare well with the results obtained after complete wet-chemical destruction[57].

Jackson and Dellar[48] attained a detection limit of 0.1 μg L$^{-1}$ total mercury in potable and natural waters using ultraviolet irradiation–atomic absorption spectrometry.

The reproducibilities for inorganic mercury and methylmercury were similar. At the 0.33 μg L$^{-1}$ level, the 95% confidence interval is approximately 0.1 μg L$^{-1}$ and at the 2.00 μg L$^{-1}$ level it is approximately 0.20 μg L$^{-1}$. The limit of detection based on the variation of results when estimating low levels is 0.1 μg L$^{-1}$ may be estimated for 95% confidence intervals to the nearest 0.1 μg L$^{-1}$ up to 0.5 μg L$^{-1}$ and over this figure to the nearest 20% (Table 6.4).

Jackson and Dellar[48] emphasise that to obtain best results for concentrations of mercury present in water the container and water itself must be stabilised to avoid loss or gain of mercury. Acidic potassium dichromate is believed to be the best preservative[58]. Any particulate matter present in a water is likely to absorb dissolved mercury.

Direct cold vapour atomic absorption spectrometry with a reduction–aeration technique gives reproducible results down to levels of about 0.5 μg L$^{-1}$. This sensitivity is not enough to monitor background levels of mercury in unpolluted areas or to meet for example, the proposed World Health Organisation requirement for total mercury in potable water of 1 μg L$^{-1}$ where it would be necessary to be able to determine mercury at one tenth of this limit (World Health Organisation[59,60]). Consequently, a preconcentration step is necessary which also separates the mercury from interfering substances. Two types of preconcentration steps have been studied, (i) those involving amalgamation with noble metals such as silver and gold and (ii) those involving preconcentration on resins.

**Table 6.4.** Recovery (%) of mercury from 150 mL of potable water sample by photolytic-cold vapour AA method. (From Jackson and Dellar[48]. Reproduced by permission of Elsevier Science Publishers Ltd.)

| Mercury present | Inorganic mercury (as nitrate) | | Methylmercury (as chloride) | | Phenylmercury (as acetate) | |
|---|---|---|---|---|---|---|
| | 0.33 μg L$^{-1}$ | 2.00 μg L$^{-1}$ | 0.33 μg L$^{-1}$ | 2.00 μg L$^{-1}$ | 0.33 μg L$^{-1}$ | 2.00 μg L$^{-1}$ |
| Potable waters | | | | | | |
| A Hardness 30 mg L$^{-1}$ | 107.106 | 102.99 | 101.85 | 99.98 | 100 | 91 |
| B Hardness 150 mg L$^{-1}$ | 98.114 | 99.102 | 95.100 | 104.105 | 93 | 94 |
| C Hardness 250 mg L$^{-1}$ | 111.96 | 99.94 | 100.97 | 97.101 | 100 | 95 |
| D Harness 358 mg L$^{-1}$ | 100.103 | 99.100 | 102.102 | 96.102 | 97 | 88 |
| Mean recovery (%) | 104 | 99 | 98 | 100 | 98 | 92 |
| Reproducibility (%) | 6.3 | 2.5 | 5.7 | 3.3 | — | — |
| Reproducibility (μg) | 0.003 | 0.007 | 0.003 | 0.010 | — | — |
| Reproducibility (μg L$^{-1}$) | 0.021 | 0.050 | 0.019 | 0.066 | — | — |
| 95% Confidence interval (μg L$^{-1}$) | 0.058 | 0.14 | 0.053 | 0.18 | — | — |

*(i) Amalgamation methods* An example of an amalgamation technique is one in which concentrated nitric acid is used to decompose organomercury compounds in water samples prior to estimation by flameless atomic absorption spectroscopy. Stannous chloride was used to liberate elementary mercury, which is then vaporised by passing a stream of air (1360 mL per min.) through the solution. The air stream passes over silver foil, where mercury is retained by amalgamation and other volatile substances pass out of the system. The foil is heated at 350°C in an induction coil, and the air stream carries the mercury vapour through a cell with quartz windows. The atomic absorption at 253.65 nm is measured and the mercury concentration (up to 0.02 µg $L^{-1}$) is determined by reference to a calibration graph. In addition to improving specificity, this method has the additional advantage of improving sensitivity due to the concentration factor achieved in collecting the mercury from a large volume a sample on the silver foil.

*(ii) Resin preconcentration* Yamagami et al.[38] applied diethiocarbamate type chelating resins to the determination of µg $L^{-1}$ of mercury in water. The samples were adjusted to pH 2.3 and passed through a column packed with 5 g of the resin, at a flow rate of 50 mL per min. The resin is then refluxed with concentrated nitric acid, and the mercury is determined by atomic absorption spectrophotometry, using the reduction–aeration technique. The method is relatively simple and inexpensive and the detection limit is 10 ng mercury in water samples as large as 10 litres i.e., 0.1 µg $L^{-1}$, thereby meeting the regulation requirements of the World Health Organisation[59,60].

Minagawa et al.[37] use dithiocarbonate resin preconcentration on a column to determine traces of inorganic and organic mercury down to 0.2 µg $L^{-1}$ in fresh water samples. River water and the other fresh waters were sampled in a 20 L high density polyethylene bottle which was rinsed three times with the water sampled before the sample was taken. The sample was adjusted to pH 2 with concentrated nitric acid, 1 mg of $HAuCl_4$ being added as preservative. Samples of water should be analysed within one week of collection to avoid losses of mercury by adsorption and vaporisation.

The apparatus used for preconcentration consisted of a column (15 mm i.d., 5 cm long) for the resin and a 20 L high density polyethylene bottle as reservoir for the samples. The 20–50 mesh wet dithiocarbamate treated resin was packed in the column. Each polyethylene bottle was cleaned by soaking in (1 + 9) nitric acid for 2 days and then rinsed throughly with distilled deionised water before use. Collected mercury is readily eluted with a slightly acidic aqueous 5% thiourea solution. The resin can then be reused.

For the determination of inorganic mercury, the 10 mL aliquot of well-mixed effluent was placed in the reaction vessel, 10 mL of 30% (w/v) potassium hydroxide was added followed by 2 mL of the tin(II) chloride solution and the air flow was started immediately. This mixture was allowed to react for 30 s, during which time the mercury vapour generated passed through the quartz gas cell in an atomic absorption spectrometer.

For the determination of total mercury, the same procedure was used, except for reduction with the tin(II) chloride Cd(II) chloride mixture (10%–1%), instead of tin(II) chloride alone. The peak heights were again measured. The total mercury minus the inorganic mercury gives an estimate of the organic mercury.

The total blank was determined by carrying out the complete procedure (using 20 L of distilled-deionised water. Five replicate measurements gave mean blanks of $0.07 \pm 0.05$ ng $L^{-1}$ and $0.15 \pm 0.06$ ng $L^{-1}$ for inorganic and total mercury, respectively.

Typical mercury contents found in unfiltered samples of Japanese rivers by this method were in the range 9 to 15 ng $L^{-1}$ with a standard deviation of about 0.4 ng $L^{-1}$, 35%–50% of which was organically bound. In excess of 90% recovery was obtained in spiking experiments in which mercuric chloride and methylmercuric chloride were added to river water samples.

Possible interferences are other ions, amino acids and naturally-occurring chelating agents which could affect the preconcentration, desorption and reduction steps. No interference was produced in the determination of 0.1 pg of mercury(II) by the presence of at least 1 ng of each of the following ions or substances added to 5 L aliquots of river water: $Cr^{3+}$, $Mg^{2+}$, $Na^+$, $K^+$, $Ca^{2+}$, $Ni^{2+}$, $Cu^{2+}$, $Pb^{2+}$, $Au^{3+}$, $Fe^{3+}$, $Al^{3+}$, $Zn^{2+}$, $PO_4^{3-}$, $Cl^-$, $CO_3^{2-}$, $NO_3^-$, $SO_4^{2-}$, silicate, cysteine and humic acid.

Oda and Ingle[39] have described a method which has a detection limit of about 5 mg $L^{-1}$ using a 1 mL sample for the determination of both forms of mercury. The inorganic and organic mercury are selectively reduced by stannous chloride and sodium borohydride and volatilised elemental mercury determined by cold vapour atomic absorption spectrometry. This procedure is much faster than other procedures because no time-consuming sample extraction, sample decomposition, or chromatographic separation steps are required.

The accuracy of organomercury determination is better than in many procedures because the organomercury concentration is not determined by difference which is difficult if most of the total mercury is inorganic mercury.

These workers showed that nitric acid is not suitable for fixing samples between sampling and analysis as it alters the ratio of organic to inorganic mercury and, indeed, causes a 25% loss of total mercury during 8 days sample storage. For extended periods of sample storage a fixing reagent comprising 10 mg $L^{-1}$ potassium chromate and 50 mL $L^{-1}$ nitric acid should be used. Even with this reagent the organic to inorganic mercury ratio will alter during storage but total mercury remains unaltered. The information on speciation can be obtained only when the sample is analysed immediately after fixing.

Ahmed and Stoeppler[49] used cold vapour atomic absorption spectrometry to carry out ultraviolet light stability studies on methylmercury in water.

## Gas chromatography

Nishi and Horimoto[28,61] determined trace amounts of methyl, ethyl and phenyl mercury compounds in river waters and industrial effluents. In this procedure the organomercury compound in 100–500 mL of sample is extracted with 0.1% L-cysteine solution, and recovered from the complex by extracting with 1 mL of benzene in the presence of hydrochloric acid and submitted to gas chromatography using a stainless-steel column (197 cm × 3 mm) packed with 5% of diethylene glycol succinate on Chromosorb W (60–80 mesh) with nitrogen as carrier gas (60 mL per minute) and an electron-capture detector.

The calibration graph is rectilinear for less than 0.1 µg of mercury compound per mL of the cysteine solution. This method is capable of determining mercury down to 0.4 µg $L^{-1}$ for the methyl and ethyl derivatives and 0.86 µg $L^{-1}$ for the phenyl derivative.

The above method has been modified by the addition of mercuric chloride to displace sulfuric bonded methylmercury groups for the determination of methylmercury(II) compounds in aqueous media containing sulfur compounds that affect the extractions of mercury. The modified method is capable of handling samples containing up to 1 µg $L^{-1}$ of various organic and inorganic sulfur compounds.

Dressman[62] used the Coleman 50 system in his determination of dialkylmercury compounds in river waters. These compounds were separated in a glass column (1.86 m × 2 mm) packed with 5% of DC-200 plus 3% of AF-1 on Gas Chrom (80–100 mesh) and temperature programmed from 70 to 180°C at 20°C $min^{-1}$, with nitrogen as carrier gas (50 mL $min^{-1}$). The mercury compound eluted from the column was burnt in a flame ionisation detactor and the resulting free mercury was detected by a Coleman Mercury Analyser MAS-50 connected to the exit of the flame ionisation instrument; down to 0.1 mg of mercury could be detected. River water (1 L) was extracted with pentane-ethyl either (4:1) (2 × 60 mL). The extract was dried over sodium sulphate, evaporated to 5 mL and analysed as above.

Ealy et al.[63] have discussed the determination of methyl, ethyl and methoxymercury halides in water and fish. The mercury compounds were separated from the samples by leaching with M-sodium iodide for 24 hours and then the alkylmercury iodides were extracted into benzene. These iodides were then determined by gas chromatography of the benzene extract on a glass column packed with 5% of cyclohexane–succinate on Anakron ABS (70–80 mesh) and operated at 200°C with nitrogen (56 mL $min^{-1}$) as carrier gas and electron capture detection. Good separation of chromatographic peaks were obtained for the mercury compounds as either chlorides, bromides or iodides. The extraction recoveries were monitored by the use of alkyl mercury compounds, labelled with 208 Hg.

Another application of gas chromatography to natural water analysis is that of Longbottom[64] who uses a Coleman 50 Mercury Analyser System, as a gas chromatographic detector. A mixture of dimethyl, diethyl-, dipropyl- and dibutylmercury (1 ng of each) was separated on a 2 m column packed with 5% of DC-200 and 3% of QF-1 on Gas-Chrom Q and temperature programmed from 60° to 180° at 20° per min. The mercury detector system was used after the column effluent had passed through a flame ionisation detector; the heights of the resulting four peaks were related to the percentages of mercury in the compounds.

Mushak et al.[65] have described a gas chromatographic method for the determination of inorganic mercury in water, urine and serum. The inorganic mercury in the sample is reacted with lithium pentafluorobenzenesulphinate arylating reagent which converts inorganic mercury to arylmercury compounds. The arylmercury compounds as well as any other organomercury compounds present in the original sample are then determined by a technique based on that described by Westoo[66,67] involving gas chromatography on columns of 10% of Dexsil-300 on Anakrom SD (70–80 mesh) and of Durapak Carbowax 400 on Porasil F (80–100 mesh). The recoveries and precision (standard deviations) were for water 70.5% (6.8), urine 81.4% (10.5) and serum 51% (9.4). The limit of detection of inorganic mercury achieved in this method was 20 µg $L^{-1}$.

Jones and Nickless[68,69] have devised methods for the determination of inorganic mercury based on conversion to an organomercury compound with arene sulphinites and formation of the trimethylsilyl derivatives, both of which are amenable to gas chromatography.

Cappon and Crispin Smith[70] have described a method for the extraction, clean-up and gas chromatographic determination of organic (alkyl and aryl) and inorganic mercury in water. Methyl-, ethyl-, and phenylmercury are first extracted as the chloride derivatives. Inorganic mercury is then isolated as methylmercury upon reaction with tetramethyltin. The initial extracts are subjected to thiosulphate clean-up and the organomercury species are isolated as the bromide derivatives. Total mercury recovery ranged between 75 and 90% for both forms of mercury, and is assessed by using appropriate 203 Hg-labelled compounds for liquid scintillation spectrometric assay. Specific gas chromatographic conditions allow detection of mercury concentrations of 1 µg $L^{-1}$ or lower. Mean deviation and relative accuracy average 3.2 and 2.2% respectively.

Zarnegar and Mushak[71] have described a gas chromatographic procedure for the determination of organomercury compounds and inorganic mercury in natural water. The sample is treated with an alkylating or arylating reagent and the organomercury chloride is extracted into benzene. Gas chromatography is carried out using electron capture detection. The best alkylating or arylating reagents were pentacyano(methyl) cobaltate(III) and tetraphenylborate. Inorganic and organic mercury could be determined sequentially by extracting and analysing two aliquots of sample, of which only one had been treated with alkylating reagent. The limits of detection achieved in the method were 10–20 ng.

Lee[72] preconcentrated methyl and ethyl mercury compounds in sub nanogram amounts from natural waters on to sulphhydryl cotton fibre. The fibre was prepared by soaking cotton for 4 to 5 days in a mixture of thioglycolic acid, acetic anhydride, acetic acid, sulphuric acid and water at 40–45°C. Adsorbed mercury compounds were eluted from the cotton fibre with hydrochloric acid/sodium chloride and extracted with benzene. Using gas chromatography with electron capture detection the detection limits were 0.04 ng $L^{-1}$ using a 20 L sample volume. Precision was approximately 20%. The method was applied to snow and freshwater samples of varying humic acid content. Methylmercury concentrations varied between 0.9 and 0.22 ng L$^{-1}$ and recoveries of spiked methylmercury varied between 42 and 68% and were strongly correlated with the concentration of humic substances. Methylmercury concentration in snow was 0.28 ng L$^{-1}$ and the recovery was 79%.

### 6.1.2 SEA WATERS

**Discussion**

Fish frequently have 80–100% of the total mercury in their bodies in the form of methylmercury, regardless of whether the sites at which they were caught were polluted with mercury or not. Methylmercury in the marine environment may originate from industrial discharges or be synthesised by natural methylation processes. Fish do not themselves methylate inorganic mercury, but can accumulate methylmercury from both sea water and food. Methylmercury has been detected in sea water only from Minamata Bay, Japan, an area with a history of gross mercury pollution from industrial discharge. It has been found in some sediments but at very low concentrations, mainly from areas of known mercury pollution[13]. It represents usually less than 1% of the total mercury in the sediment and frequently less than 0.1%[13]. Microorganisms within the sediments are considered to be responsible for the methylation[13], and it has been suggested that methylmercury may be released by the sediments to the sea water, either in dissolved form or attached to particulate material and thereafter rapidly taken up by organisms.

Compeau and Bartha[75] studied the effects of sea salt anions on the formation and stability of methylmercury. The effect of different anions in sea water on the formation and stability of methylmercury was investigated. The extent of methylation was reduced in the presence of sulphide under anaerobic conditions and of bicarbonate under both aerobic and anaerobic conditions; other anions had no significant effect. In the dark monomethylmercuric chloride was chemically stable in the presence of all the anions tested.

Davies et al.[74] set out to determine the concentrations of methylmercury in sea water samples much less polluted than Minamata Bay, viz the Firth of Forth, Scotland. They described a tentative bioassay method for determining methylmercury at the 0.06 µg g$^{-1}$ level. Mussels from a clean environment were suspended in cages. A small number were removed periodically, homogenised and analysed for methylmercury. The rate of accumulation of methylmercury was determined, and by dividing this by mussel filtration rate, the total concentration of methylmercury in the sea water was calculated. The mussels were extracted with benzene and the extract analysed by gas chromatography.

The methylmercury concentration in caged mussels increased from low levels (less than 0.01 µg g$^{-1}$) to 0.06–0.08 µg g$^{-1}$ in 150 days giving a mean uptake rate of 0.4 ng $g^{-1}$ $d^{-1}$. The average percentage of total mercury in the form of methylmercury increased from less than 10% after 20 days to 33% after 150 days.

Davies et al.[74], calculated that the total methylmercury concentration in the sea water as 60 pg $L^{-1}$ (i.e. 0.1–0.3% of the total mercury concentration), as opposed to 32 ng $L^{-1}$ methylmercury found in Minamata Bay, Japan.

Stoeppler and Matthas[76] have made a detailed study of the storage behaviour of methylmercury and mercuric chloride in seawater. They recommend that samples spiked with inorganic and/or methylmercury chloride be stored carefully in cleaned glass containers acidified with hydrochloric acid to pH 2.5. Brown glass bottles are preferred. Storage of methylmercury chloride should not exceed 10 days.

Olsen[78] has reported that up to 20% losses of mercury occur during storage of seawater samples for up to 96 h prior to analysis. They attributed losses to either a salting out effect which increased the volatilisation of mercury or biochemical reactions which either increase the volatility of or precipitate mercury.

**Subtractive differential pulse voltammetry**

Sipso et al.[77] used this technique, utilising a twin gold electrode to determine mercury levels in seawater samples taken in the North Sea.

**Atomic absorption spectrometry**

Fitzgerald and Lyons[45] have described flameless atomic absorption methods for determining organic mercury compounds in coastal and seawaters. These workers used ultraviolet light in the presence of nitric acid to decompose the organomercury compounds. In this method two sets of 100 mL samples of natural water are collected in glass bottles and then adjusted to pH 1.0 with nitric acid. One set of samples is analysed directly to give inorganically bound mercury, the other set is photo-oxidised by means of ultraviolet radiation to destroy organic material and then analysed to give total mercury. The element is determined by a flameless-atomic-absorption technique, after having been collected on a column of 1.5% of OV-17 and 1.95% of QF-1 on Chromosorb W-HP (80–100 mesh), cooled in liquid nitrogen bath and then released by heating the column. The precision of analysis is 15%. It was found that up to about 50% of the mercury present in river and coastal waters was organically bound or associated with organic matter.

A method developed by Dean and Rues[79] is suitable for the determination of 10–100 ng $L^{-1}$ dissolved inorganic mercury and those organomercury compounds which form dithizonates in saline, sea and estuary waters. In this method, inorganic mercury is extracted from the acidified saline water as its dithizonate into carbon tetrachloride. Organomercury compounds was also be extracted by the carbon tetrachloride, but not all these compounds form dithizonates and those which do not (e.g. dialkylmercury compounds) may not be determined by this method. Mercury is recovered from the organic extracts by back extraction with acid and determined by stannous chloride reduction followed by cold vapour atomic absorption spectrometry.

The performance characteristics of this method are tabulated in Table 6.5.

Millward and Bihan[80] studied the effect of humic material on the determination of mercury by flameless atomic absorption spectrometry. In both sea and fresh water, association between inorganic and organic entities takes place within 90 min at pH values of 7 or above, and, consequently, the organically bound mercury is not detected by an analytical method designed for inorganic mercury. The amount of detectable mercury was related to the amount of humic material added to the solutions. However, total mercury could be measured after exposure to ultraviolet radiation under strongly acid conditions.

Agemian and Chau[81] showed that organomercurials could be quantitatively decomposed by ultraviolet radiation and that the rate of decomposition of organomercurials increased rapidly in the presence of sulphuric acid and with increased surface area of the ultraviolet irradiation. They developed a flow-through ultraviolet digestor which had a delay time of 3 min to carry out the photooxidation in the automated system. The ultraviolet radiation has no effect on chloride. The method, therefore, can be applied to both fresh and saline waters without the chloride interference. With an atomic absorption spectrometric finish this method was capable of determining down to 0.02 µg $L^{-1}$ mercury.

Yamagami et al.[38] evolved a technique involving amalgamation of methylmercury in sea water on to gold followed by atomic absorption spectrophotometry for the determination of picogram quantities of the organomercury compound.

Table 6.5. Department of the Environment (UK) conditions for determination of organomercury compounds.

| | |
|---|---|
| Range of application | Up to 100 ng $L^{-1}$ |
| Calibration curve | Linear to 250 ng $L^{-1}$ |
| Standard deviation | Mercury concentration (ng $L^{-1}$)     Standard deviation (ng $L^{-1}$) <br> 0.0     1.30 <br> 50.0     1.15 <br> (each with 9 degrees of freedom) |
| Limit of detection | 4 ng $L^{-1}$ (with 9 degrees of freedom) |
| Sensitivity | 100 ng $L^{-1}$ is equivalent to an absorbance of approximately 0.1 |
| Bias | None detected |
| Interferences | The combined effect of the commonly presently ions in estuarine and sea waters at the concentration normally encountered in these waters is less than 1 ng $L^{-1}$ at a mercury concentration of 30 ng $L^{-1}$ |
| Time required for analysis | For 6 samples the total analytical and operator times are approximately 140 minutes and 60 minutes respectively |

Samples of sea water, ground-water and river water were analysed for methylmercury and total mercury. Methylmercury is extracted with benzene and concentrated by a succession of three partitions between benzene and cysteine solution. Total mercury is extracted by wet combustion of the sample with sulphuric acid and potassium permanganate. The proportion of methylmercury to total mercury in the coastal sea water sampled was around 1%.

Graphite furnace atomic absorption spectrophotometry has also been applied to the determination of trace levels of divalent mercury in inorganic and organomercury compounds in sea water. Filippelli[82] has described a technique in which mercury is first preconcentrated using the ammonium tetramethylenedithiocarbamate (ammonium pyrrolidine-dithiocarbamate)–chloroform system and then determined by graphite furnace atomic-absorption spectrometry. The technique is capable of detecting mercury(II) in the range 5–1500 ng in 2.5 mL of chloroform extract and can be adapted to detect subnanogram levels. Atmospheric pressure helium microwave-induced plasma emission spectrometry has been used as an element-selective detector for gas chromatography of organomercury compounds in sea water[83].

Chiba et al.[84] used atmospheric pressure helium microwave induced plasma emission spectrometry with the cold vapour generation technique combined with gas chromatography for the determination of methylmercury chloride, ethylmercury chloride and dimethylmercury in sea water following a 500 fold preconcentration using a benzene–cysteine extraction technique.

The analysis system consisted of a Shimsdzu QC-6A gas chromatograph, a chemically deactivated four-way valve for solvent ventilation, a heated transfer tube interface, a Beenakker-type $TM_{010}$ microwave resonance cavity, and an Ebert-type monochromator (0.5 m focal length).

The dual column gas chromatograph was equipped with a thermal conductivity detector. The interface between the gas chromatograph and the discharge tube of the microwave-induced plasma detector is constructed from a chemically deactivated four-way valve and a heated transfer tube. The gas chromatographic columns and optimum operating conditions are summarised in Table 6.6. As is seen in Table 6.6 the diethylene glycol succinate column was used for the measurement of methylmercury chloride and ethylmercury chloride and the OV-17 column for that of dimethylmercury. The former column was treated with dimethylsilane and potassium bromide in order to deactivate the surface.

The $TM_{010}$ microwave cavity is constructed from pure copper metal. The microwave generator, which provides 20–200 W of microwave power at 2.45 GHz, is run at 75 W forward power. The width and the height of both monochromator entrance and exit slits are 10 µm and 1 mm, respectively. A photomultiplier tube with low dark current and high gain over a wide wavelength region is used as a detector. The measurement of mercury is carried out at 253.7 nm mercury line.

The flow rate of carrier helium gas is adjusted at 80 mL min$^{-1}$ for both columns, and then the plasma ignited. About 30 min later, the plasma and the

**Table 6.6.** Operating conditions for gas Chromotography (From Chiba et al.[84] Reproduced by permission of American Chemical Society.)

| | Conditions for determination of | |
|---|---|---|
| | $CH_3HgCl$ and $C_2H_5HgCl$ | $(CH_3)_2Hg$ |
| Column | Pyrex, 1 m × 3 mm i.d. | Pyrex 3 m × 3 mm i.d. |
| Column packing | 15% DEGS[a] on 80/100 mesh Chromosorb W | 3% OV-17 on 80/100 mesh Uniport HP |
| Column temp. °C. | 160 | 70 |
| Injector temp. °C. | 180 | 130 |
| Detector oven temp. °C. | 180 | 130 |
| Transfer tube temp. °C. | 190 | 140 |
| Carrier gas | helium | helium |
| Carrier gas flow Rate mL min$^{-1}$ | 80 | 80 |
| Detector | Ratharometer | Ratharometer. |

[a]DEGS = diethylene glycol succinate.

temperature of the gas chromatograph stabilise. The adjustments of wavelength and observation position in the plasma are performed as follows. An appropriate concentration of methylmercury chloride standard solution (1-2 µL) is injected very slowly into the OV-17 column, and a broad mercury peak appears. During the appearance of the peak, wavelength and observation position (both vertical and horizontal) are adjusted quickly.

The chromatograms were detected with the thermal conductivity and the microwave-induced plasma detectors in series. When the microwave-induced plasma detector was used as a detector, the emission signals were monitored at 253.7 nm, and the solvent was vented through a four-way valve before reaching the microwave induced plasma detector.

The column packed with OV-17 used for the measurement of dimethylmercury was optimised in terms of carrier gas flow rate and column temperature.

Detection limit and standard deviation date obtained for these organomercury compounds without preconcentration are shown in Table 6.7.

The total extraction efficiency involved in the three stages of the extraction of methylmercury chloride from sea water by the cysteine–benzene extraction technique was reproducible at 42% for a 500 fold concentration giving a detection limit of 0.4 ng L$^{-1}$ and a relative standard deviation of 6% at the 20 ng L$^{-1}$ level.

Summarising, it can be seen that in natural and sea waters organomercury compounds can be speciated and determined in amounts down to 10 ng L$^{-1}$, thereby meeting even the most searching present day requirements.

Yamamoto et al.[85] described a procedure for the determination of methyl mercury compounds in seawater involving extraction with benzene, amalgamation of the mercury on to gold foil, then release of the mercury from

**Table 6.7.** Analytical figures of merit in the determination of alkylmercury compounds by the GC/MIP System. (From Chiba et al.[84] Reproduced by permission of American Chemical Society.)

| Compound | Detection limit ($\mu g\ L^{-1}$) | rel std dev[a] (%) | Dynamic range (decades) |
|---|---|---|---|
| Methylmercury chloride | 0.09 (0.02 $pg^{-1}/s$) | 2.0 | 5 |
| Ethylmercury chloride | 0.12 (0.02 $pg^{-1}/s$) | 2.0 | 5 |
| Dimethylmercury | 0.40 (0.03 $pg^{-1}/s$) | 3.0 | 4.5 |

[a] Measured with 1 $\mu g\ L^{-1}$ of mercury for each compound.

the foil followed by atomic absorption spectrometry. Down to 20 $\mu g\ L^{-1}$ of organomercury compounds can be determined by this technique.

Workers at the Department of Environment UK[86] have developed a similar technique which involves a preliminary extraction of the dimethylmercury compound with a carbon tetrachloride solution of dithizone.

### 6.1.3 RAINWATER

**Atomic absorption spectrometry**

Ahmed et al.[87] preconcentrated methyl mercury and inorganic mercury from rainwater on an anion exchange column prior to analysis by cold vapour atomic absorption spectrometry. Methylmercury was determined by passing a 500 mL sample (in 5% hydrochloric acid) through an anion exchange column on which ionic mercury was retained. Methylmercury passing the column was decomposed by ultraviolet radiation and the methylmercury concentration determined by difference. Recovery of ionic mercury increased as acid concentration increased with ultraviolet irradiation.

### 6.1.4 INDUSTRIAL EFFLUENTS AND WASTEWATERS

**Thin layer chromatography**

Thin layer chromatography has been used to evaluate[88] organomercury compounds in industrial effluents. $C_1-C_6$ n-alkylmercury chlorides were separated on layers prepared with silica gel (27.75 g) plus sodium chloride (2.35 g in 60 mL water) using as development solvent cyclohexane–acetone–28%aqueous ammonia (60:40:1). The $R_F$ value decrease with increasing carbon chain length and phenylmercury acetate migrated between the $C_1$ and $C_2$ compounds. The spots were detected by spraying with dithizone solution in chloroform. Water samples (100–200 mL) were treated with hydrochloric acid (to produce a concentration of 0.1–0.2 M) and with potassium permanganate solution until a pink colour persists, then shaken (x3) with chloroform (one third the volume of the aqueous layer) for 3 min. The combined extracts are shaken with 0.1–0.2 M aqueous ammonia; the aqueous solution is neutralised to *p*-nitrophenol with dilute hydrochloric acid and adjusted to 2 M in hydrochloric acid and the organomercury compounds extracted with chloroform. The chloroform extract usually recovered about 95% of the organomercury compounds present in the sample and was in suitable form for thin-layer chromatography.

### Spectrophotometric methods

Itsuki and Komuro[89] determined organomercury compounds in waste water by heating the sample with 2:1 mixture of nitric and hydrochloric acids (10 mL) and 30% hydrogen peroxide (2 mL) at 90°C for one hour followed by the addition of 50% ammonium citrate solution (5 mL), diaminocyclohexane–tetraacetone (5% v/v in 2% sodium hydroxide) (5 mL) and 10% hydroxylamine hydrochloride solution (1 mL) followed by pH adjustment to pH 3–4 with aqueous ammonia. The solution is shaken with 5 mg mL$^{-1}$ 1, 1, 1-trifluoro-4-(2-thienyl)-4-mercaptabuta-3-en-2-one in benzene (10 mL) the benzene layer washed with 0.1 M borate pH 11 (50 mL) and the solution evaluated spectrophotometrically at 365 nm.

Murakami and Yoshinaga[90] determined organomercury compounds in industrial effluents by a spectrophotometric procedure using dithizone The sample (100 mL) is neutralised to *p*-nitrophenol and hydrochloric acid added to give an acid concentration of 0.1–0.2 M. The solution is shaken with chloroform (one third the volume of the sample), 6% potassium permanganate solution is added (until the mixture is pink) then 0.2 mL in excess, the mixture is shaken for 3 min, and the chloroform phase removed. 10% hydroxylammonium chloride solution is added to decolorise the aqueous phase and the extraction repeated twice with chloroform. The combined chloroform extracts are washed with 0.1 M hydrochloric acid then the mercury is extracted with 0.1–0.5 M aqueous ammonia, the aqueous solution is filtered and mercury determined.

### Atomic absorption spectrometry

Carpenter[91] has reviewed the application of flameless atomic absorption spectroscopy to the determination of mercury in paper mill effluents. Thiosulphate oxidation is recommended as a means of converting organomercury compounds to inorganic mercury. Carpenter concludes that the practical limit of detection of mercury in effluents using this technique is 1 µg L$^{-1}$.

### Gas chromatography

The Nishi and Horimoto[28,61] procedure described in Section 6.1.1 (gas chromatography) has been applied to the determination of methyl, ethyl and phenyl organomercury compounds in industrial effluents.

## 6.1.5 SEWAGE EFFLUENTS

**Thin layer chromatography**

Takeshita[92] has used thin-layer chromatography to detect alkylmercury compounds and inorganic mercury in sewage. The dithizonates were prepared by mixing a benzene solution of the alkylmercury compounds and a 0.4% solution of dithizone. When a green coloration was obtained the solution was shaken with M sulphuric acid followed by aqueous ammonia and washed with water. The benzene solution was evaporated under reduced pressure, and the dithizonates, dissolved in benzene, were separated by reversed phase chromatography on layers of corn starch and Avicel SF containing various proportions of liquid paraffin. Solutions of ethanol and 2-methoxyethanol were used as developing solvents. The spots were observed in daylight. The detection limits was from 5 to 57 ng (calculated as organomercury chloride) per spot.

**Atomic absorption spectrometry**

The Department of the Environment (UK)[54] has described a method for determining total organic plus inorganic mercury in non-saline sewage and effluents. All forms of mercury are converted to inorganic mercury using prolonged oxidation with potassium permanganate[93]. Solid samples require a more prolonged and vigorous oxidation to bring the mercury completely into solution in the inorganic form. A modification of the Uthe digestion procedure is used for such samples[94]. The inorganic mercury is determined by the flameless atomic absorption spectrophotometric technique using a method similar to that described by Osland[95]. Acid stannous chloride is added to the sample to produce elemental mercury:

$$Hg^{2+} + Sn^{2+} \longrightarrow Hg^0 + Sn^{4+}$$

The mercury vapour is carried by a stream of air or oxygen into a gas curvette placed in the path of the radiation from a mercury hollow cathode lamp and the absorption of this radiation at 253.7 nm by the mercury vapour is measured. Many of the potential interferences in the atomic absorption procedure are removed by the preliminary digestion/oxidation procedure. The most significant group of interfering substances is volatile organic compounds which absorb radiation in the ultraviolet. Most of these are removed by the pretreatment procedure used and the effect of any that remain are overcome by pre-aeration. Bromide and iodide ions may cause interference. Substances which are reduced to the elemental state by stannous chloride and then form a stable compound with mercury may cause interference, e.g. selenium, gold, palladium and platinum. The effects of various anions, including bromide and iodide, were studied. These are not likely to be important interferers. Excellent performance characteristics are presented for this method (see Table 6.8).

**Table 6.8.** Performance characteristics WRC method for mercury.

| | |
|---|---|
| Range of application | Up to 2.0 µg L$^{-1}$ for liquid samples. Up to 2.0 µg g$^{-1}$ for solid samples. |
| Calibration curve | Linear to 20 µg L$^{-1}$ for liquid samples. Linear to 20 µg g$^{-1}$ for solid samples. |
| Within-batch standard deviation for liquid samples | Mercury concentration (µg L$^{-1}$) / Standard deviation (µg L$^{-1}$): 0.0 / 0.025; 0.2 / 0.022; 2.0 / 0.024; 0.0 / 0.078; 0.2 / 0.032; 2.0 / 0.104 (All with 9 degrees of freedom) |
| for solid samples | Not known |
| Limit of detection | 0.1–0.2 µg L$^{-1}$ (with 9 degrees of freedom) for liquid samples depending on their nature. Not known for solid samples. |
| Sensitivity | 10 µg L$^{-1}$ for liquid samples and 10 µg g$^{-1}$ for solid samples are equivalent to an absorbance of approximately 0.3. |

### Miscellaneous

Van Ettekoven[96] has described a direct semi-automatic scheme based on ultraviolet light absorption for the determination of total mercury in water and sewage sludge. The full determination time is about 10 min. The lower limit of detection of mercury in water is 0.03 µg L$^{-1}$ and 0.2 mg L$^{-1}$ (mg kg$^{-1}$ dry matter) in sewage sludge.

### 6.1.6 SOILS AND RIVER SEDIMENTS

In lakes, streams and rivers, mercury can collect in the bottom sediments, where it may remain for a long time. It is difficult to release this mercury from the matrices for analysis.

Mercury is also found in soil as a result of applications of mercury containing compounds, or sewage contaminated with traces of mercury.

### Spectrophotometric procedures

Kimura and Miller[3] described a procedure for the determination of organomercury, (methylmercury, ethylmercury and phenylmercury compounds) and inorganic mercury in soil. In this method, the sample is digested in a steam bath with sulfuric acid (1.8 N) containing hydroxy-ammonium sulphate, sodium chloride and, if high concentrations of organic matter are present, potassium sulfate solution. Then 50% hydrogen peroxide is added in portions with vigorous mixing until the solution becomes blue-green to yellow in colour, then heating is continued to decompose excess hydrogen peroxide. Potassium permanganate is added until the pink colour persists for 15 min then the cooled solution reduced with hydroxy-ammonium sulfate dissolved in sodium chloride. Air is then passed through this solution to sweep out elemental mercury into an absorber containing potassium permanganate in dilute sulfuric acid, and stannous chloride. Following reduction with hydroxy-ammonium sulphate and acidification with dilute sulfuric acid, mercury is determined in a chloroformic dithizone extract of this solution spectrophotometrically at 605 nm.

In a similar procedure[97] the sediment is wet oxidised with dilute sulfuric acid and nitric acids in an apparatus in which the vapour from the digestion is condensed into a reservoir from which it can be collected or returned to the digestion flask as required. The combined oxidised residue and condensate are diluted until the acid concentration is 1 N and nitrate is removed by addition of hydroxylammonium chloride with boiling. Fat is removed from the cooled solution with carbon tetrachlodithizone in carbon tetrachloride. The extract is shaken with 0.1 N hydrochloric acid and sodium nitrite solution and, after treatment of the separated aqueous layer with hydroxylammonium chloride a solution of urea and then EDTA solution are added to prevent subsequent extraction of copper. The liquid is then extracted with a 0.01% solution of dithizone in carbon tetrachloride and mercury estimated in the extract spectrophotometrically at 485 nm.

Kimura and Miller[98] also described the following methods for the determination in soil samples of extractable organic mercury, total mercury and extractable ionic mercury.

### Extractable phenyl- and alkylmercury compounds

Phenyl and alkylmercury compounds are extracted from about 1 g soil by shaking for 2 h with 25 mL 0.1 M phosphate pH 8 buffer containing 6 mg thiomalic acid, added just prior to use, and analysed after dilution of a 5 mL aliquot of the centrifuged extract with 5 mL water, and acidification with 5 mL 9 N hydrochloric acid containing 150 mg hydroxylammonium chloride. The final determination is made by the dithizone microprocedure of Miller and Polley[99]. Diphenyl- and dialkylmercury compounds are extracted from 1 g soil by shaking for 2 h with 10 mL chloroform and analysed by cleaving the disubstituted mercurial to give an aryl- or alkylmercury salt, using 9 N or 12 N hydrochloric acid, followed by the dithizone microprocedure[99].

### Ionic mercury

Ionic mercury is extracted from about 1 g soil by shaking for 2 h with each of two 25 mL portions of 2 M sodium chloride. The combined centrifuged and filtered (using 1 M sodium chloride for washing) extract is analysed by the procedure of Polley and Miller[99].

**Total mercury**

Total mercury is determined in soils containing phenylmercury acetate and or ethylmercury acetate using the method described by Polley and Miller[99]. Total mercury is determined in soils containing methylmercury chloride and methylmercury dicyanamide by the method described by Kimura and Miller[3].

Kimura and Miller[3] have also studied the decomposition of organic fungicides in soil to mercury vapour and to methyl- or ethylmercury compounds and devised methods for the determination of these compounds in the vapours liberated from the soil sample. The mixed vapours of mercury and organomercury compounds is passed successively through bubblers containing a sodium carbonate–diabasic sodium phosphate solution which absorbs 45–99% of organic mercury and through an acidic potassium permanganate solution to absorb inorganic mercury vapour. In both cases the mercury in the scrubber solution is determined photometrically at 605 nm with dithizone.

**Alternate digestion procedures prior to spectroscopy**

A disadvantage of all these procedures is that the lowest concentration of mercury that can be determined in the soil or sediment samples is of the order of $0.05-1$ mg kg$^{-1}$. These high detection limits are in part due to high blanks caused by the multiplicity of digestion reagents used in the procedures. Several investigators have liberated mercury from soil and sediment samples by application of heat to the samples and collection of the released mercury on gold surfaces. The mercury was then released from the gold by application of heat or by absorption in a solution containing oxidising agents[100].

Bretthaur et al.[101] and Anderson et al.[102] described a method in which samples were ignited in a high-pressure oxygen-filled bomb. After ignition, the mercury was absorbed in a nitric acid solution. Pillay et al.[103] used a wet-ashing procedure with sulfuric acid and perchloric acid to digest samples. The released mercury was precipitated as the sulfide. The precipitate was then redigested using aqua regia.

Feldman[58] digested solid samples with potassium dichromate, nitric acid, perchloric acid and sulfuric acid. Bishop et al.[104] used aqua regia and potassium permanganate for digestion.

The approved U.S. Environmental Protection Agency[105] digestion procedure requires aqua regia and potassium permanganate as oxidants.

These digestion procedures are slow and often hazardous because of the combination of strong oxidizing agents and high temperatures. In some of the methods, mercuric sulfide is not adequately recovered. The oxidising reagents, especially the potassium permanganate, are commonly contaminated with mercury, which prevents accurate results at low concentrations.

## Atomic absorption spectrophotometry

Earlier work on the determination of total mercury in river sediments includes that of Iskandor et al.[106]. Iskandor applied flameless atomic absorption to a sulfuric acid nitric acid digest of the sample following reduction with potassium permanganate, potassium persulfate and stannous chloride. A detection limit of 1 µg kg$^{-1}$ is claimed for this somewhat laborious method. Langmyhr and Aamodt[107] determined down to 0.1 µg L$^{-1}$ of organomercury and Matsunaga and Takahashi[108], Craig and Mortan[109] and the AOAC[110] determined organic mercury in river sediments using cold vapour atomic absorption spectrometry.

A method[111] has been described for the determination of down to 2.5 µg kg$^{-1}$ alkylmercury compounds and inorganic mercury in river sediments. This method uses steam distillation to separate methylmercury in the distillate and inorganic mercury in the residue. The methylmercury is then determined by flameless atomic absorption spectrophotometry and the inorganic mercury by the same technique after wet digestion with nitric acid and potassium permanganate. The well known adsorptive properties of clays for alkylmercury compounds does not cause a problem in the above method. The presence of humic acid in the sediment did not depress the recovery of alkylmercury compounds by more than 20%. In the presence of metallic sulfides in the sediment sample the recovery of alkylmercury compounds decreased when more than 1 mg of sulfur was present in the distillate. The addition of 4 N hydrochloric acid, instead of 2 N hydrochloric acid before distillation completely, eliminated this effect giving a recovery of 90–100%.

This excellent method was sufficiently sensitive to determine 0.02 mg kg$^{-1}$ methyl mercury and 9 mg kg$^{-1}$ inorganic mercury in river sediment samples.

Jurka and Carter[112] have described an automated determination of down to 0.1 mg kg$^{-1}$ total mercury in river sediment samples with a precision of 0.13 to 0.21 µg Hg kg$^{-1}$ at the 1 mg Hg kg$^{-1}$ level and with standard deviations varying from 0.011 to 0.02 mg Hg kg$^{-1}$ (i.e. relative standard deviations of 8.4 to 12%). At the 17.2 to 32.3 mg Hg kg$^{-1}$ level in sediments recoveries in methyl mercuric chloride spiking studies were between 85 and 125%. This method is based on the automated procedure of El Awady et al.[6] for the determination of total mercury in waters and wastewaters in which potassium persulfate and sulphuric acid were used to digest samples for analysis by the cold-vapour technique. These workers proved that the use of potassium permanganate as an additional oxidising agent was unnecessary.

Aromatic organic compounds such as benzene, which are not oxidised in the digestion, absorb at the same wavelength as mercury. This represents a positive interference in all cold vapour methods for the determination of mercury. For samples containing aromatics (i.e. those contaminated by some industrial wastes), a blank analysis must be performed and the blank results must be subtracted from the sample results. The blank analysis is accomplished by replacing the potassium persulfate reagent and the stannous chloride reagent with distilled water and reanalysing the sample.

Umezaki and Iwamoto[7] have reported that organic mercury can be reduced directly with stannous chloride in the presence of sodium hydroxide and copper (II). The determination of organic mercury can be simplified, particularly if the reagent used for back extraction does not interfere with the reduction of organic mercury. Matsumaya and Takahasi[113] found that back extraction with an ammoniacal glutathione solution was satisfactory. In this method, contamination only from the ammoniacal glutathione solution is expected. However, any inorganic mercury in this solution will be adsorbed on the glass container walls with a half-life about 2 d (i.e. the blank value becomes zero if the solution is left to stand for more than a week). This method for mercury in sediments does not distinguish between the different forms of organomercury. Down to 0.2 µg kg$^{-1}$ mercury in sediments can be determined by this method with a standard deviation of 0.03 µg kg$^{-1}$. In this method, a large weight sample (10–20 g) is extracted with hydrochloric acid for two days and organic mercury then extracted from the filtrate with benzene. Mercury is back extracted from the benzene with aqueous ammoniacal glutathione. This extract is then added to aqueous solution containing sodium hydroxide, cupric sulfate and stannous chloride and the elemental mercury released is swept off with nitrogen and, in a further concentration step is collected on gold granules. Finally, the granules are heated at 500°C to release mercury which is determined by flameless atomic absorption spectrophotometry at 253.7 nm.

Workers at the Department of the Environment, UK[86] have described a procedure for the determination of methylmercury compounds in soils and sediments which involves extraction with a carbon tetrachloride solution of dithizone, reduction to elemental mercury then analysis by atomic absorption spectrometry.

## Gas chromatography

Bartlett et al.[114] used the method of Uthe et al.[94,115,116] for determining methylmercury. Sediment samples of 2–5 g were extracted with toluene after treatment with copper sulphate and an acidic solution of potassium bromide. Methylmercury was then back extracted into aqueous sodium thiosulphate. This was then treated with acidic potassium bromide and copper sulphate following which the methylmercury was extracted onto pesticide grade benzene containing approximately 100 µg L$^{-1}$ of ethyl mercuric chloride as an internal standard. The extract was analysed by electron capture gas chromatography using a Pye 104 chromatography equipped with 65Ni detector. The glass column (1 m × 0.4 cm) was packed with 5% neopentyl glycol adipate on Chromosorb G (AW-DMCS). Methylmercury was measured by comparing the peak heights with standards of methylmercuric chloride made up in the ethylmercury benzene solution (see Table 6.9) The detection limit was 1–2 µg kg$^{-1}$.

Ealy et al.[63] determined methyl, ethyl and methyloxyethyl mercury compounds in sediments by leaching the sample with sodium iodide for 24 hours and then extracting the alkylmercury iodides into benzene. These iodides are then

**Table 6.9.** Typical GLC retention times for organomercury dithizonates. (1) 2% polethyleneglycol succinate on Chromosorb G (acid-washed, DMCS-treated, 60–80 mesh) in glass column 1.5m long, 3mm I.D.; carrier gas, nitrogen. (From Bartlett et al.[114] Reproduced by permission of McMillan Magazines, London.)

| Dithizonate | Column temperature (°C) | | | | |
|---|---|---|---|---|---|
| | 140 | 150 | 160 | 170 | 180 |
| Methylmercury | 3.8 | 2.8 | 2.2 | 1.6 | 1.2 |
| Ethylmercury | 6.6 | 4.6 | 3.6 | 2.7 | 2.0 |
| Ethoxyethylmercury | 17.0 | 11.6 | 8.7 | 6.2 | 4.9 |
| Methoxyethylmercury | 17.4 | 12.0 | 8.7 | 6.2 | 4.9 |
| Tolylmercury | — | — | — | 29.0 | 19.5 |
| Phenylmercury | — | — | — | 42.0 | 27.0 |

determined by gas chromatography of the benzene extract on a glass column packed with 5% cyclohexylenedimethanol succinate on Anakron ABS (70 to 80 mesh) operated at 200°C with nitrogen (56 mL min$^{-1}$) as carrier gas and electron capture detection ($^3$H foil). Good separation of chromatographic peaks is obtained for the mercury compounds as either chlorides, bromides or iodides.

Bartlett et al.[114] and Longbottom et al.[117] have described a solvent extraction gas chromatographic method for the determination of methyl mercury compounds in river sediments. They observed unexpected behaviour of methylmercury containing River Mersey sediments during storage. They experienced difficulty in obtaining consistent methylmercury values; supposedly identical samples analysed at intervals of a few days gave markedly different results. They therefore followed the levels of methylmercury in selected sediments over a period of time to determine if any change was occurring on storage. They found that the amounts of methylmercury observed in the stored sediments did not remain constant; initially there is a rise in the amount of methylmercury observed and then, after about ten days, the amount present begins to decline to levels which in general only approximate to those originally present. They have observed this phenomenon in nearly all of the Mersey sediment samples they examined. It was noted that sediments sterilised, normally by autoclaving at approximately 120°C, did not produce methylmercury on incubation with inorganic mercury suggesting a microbiological origin for the methylmercury. A control experiment was carried out in which identical samples were collected and homogenised. Some of the samples were sterilised by treatment with 40 g L$^{-1}$ solution of formaldehyde. Several samples for both sterilised and unsterilised sediments were analysed at intervals and all of the samples were stored in ambient room temperature (18°C) in the laboratory. There is a difference in the behaviour between the sterilised and unsterilised samples. This work suggests that the application of laboratory derived results directly to natural conditions could, in these cases, be misleading; analytical results for day 10 if extrapolated directly might lead to the conclusion that natural methylmercury levels and rates of methylation are much greater than in fact they tally are. Work in this area with model or laboratory systems needs to be interpreted with particular caution.

Cappon and Crispin Smith[70] determined down to 0.001 mg kg$^{-1}$ alkyl and aryl mercury compounds in sediments by a procedure involving conversion to chloro derivatives followed by gas chromatography.

**Nuclear magnetic resonance spectroscopy**

Robert and Robenstein[17] carried out indirect determination of Hg$^{119}$ NMR spectra of methymercury complexes, e.g. CH$_3$Hg$^{11}$ thiol ligands in sediment samples.

### 6.1.7 PLANTS, CROPS AND AQUATIC ORGANISMS

**Spectroscopic method**

Gutermann and Lisk[119] suggested a method of overcoming mercury losses during decomposition of mercury containing organic materials by adopting the Schöniger oxygen flask combustion technique. They determined mercury in apples by first drying the apple tissue on cellophane, *in vacuo*, overnight. The dry material is then combusted in an oxygen-filled flask and the combustion products absorbed in 0.1 N hydrochloric acid. Mercury is extracted with dithizone and determined spectrophotometrically. Recovery of 0.3–0.6 mg Hg kg$^{-1}$ in apples by this procedure averaged 83.6%.

**Atomic absorption spectrometry**

Langmyhr and Aamodt[107] have described a cold vapour atomic absorption spectrometric method for the determination of nanogram amounts of mercury in aquatic organisms.

**Gas chromatography**

Ealy *et al.*[163] have discussed the determination of methyl, ethyl and methoxyethyl mercury(II) halides in environmental samples such as aquatic systems, seeds and fish. The mercury compounds were separated from the samples by leaching with M sodium iodide for 24 hours and then the alkylmercury iodides were extracted into benzene. These iodides were then determined by gas chromatography of the benzene extract on a glass column packed with 5% of cyclohexylene succinate on Anakrom ABS (70–80 mesh) and operated at 200°C with nitrogen (56 mL min$^{-1}$) as carrier gas and electron-capture detection. Good separation of chromatographic peaks were obtained for the mercury compounds as either chlorides, bromides or iodides. The extraction recoveries were monitored by the use of alkylmercury compounds labelled with[208] Hg.

Houpt and Compaan[120] used emission spectrographic analysis for the identification of traces of organic matter containing halogens, phosphorus, sulfur and mercury isolated from eggs and grass by gas chromatography. They transferred the gas chromatographic fractions sequentially, through a heated stainless-steel capillary tube, to a silica tube (3 mm i.d.) in which they were submitted to a plasma discharge (2.45 MHz) in helium at 10 Torr. The emission spectrum arising from the fragmentation, ionization and excitation of the organic molecule was then analysed with the aid of two monochromators, the intensities of the required analytical lines being measured photoelectrically. One monochromator was focused on a characteristic line (e.g., the 247.86 nm carbon line) as a chromatographic detector and, when the intensity of this line is a maximum for any one fraction detected in the discharge-tube, a 10 s sweep over the range 200–600 mm was made by the other monochromator. Examination of the resulting complete spectrograms revealed the presence or absence of mercury, phosphorus, sulphur, chlorine, bromine, iodine. This method permits the determination of 5 pg of methylmercury in plant and biological samples.

## 6.1.8 BIOLOGICAL MATERIALS

Earlier classical procedures are generally too insensitive and laborious and lack the specificity required to distinguish between different types of organomercury compounds and in many instances, do not even distinguish between organic mercury and inorganic mercury. Thus, in a typical method for determining mercury in biological tissue the tissue is digested with nitric and sulfuric acids and following the addition of sodium, acetate and formalin the solution is electrolysed at 80°C using a zinc anode and graphite on copper cathode to collect the mercury. Following stripping of the mercury from the cathode with concentrated nitric acid, addition of slight excess of potassium permanganate and decomposition of excess permanganate with hydrogen peroxide the solution is treated with an excess of standard ferric alum solution:

$$Hg^+ + Fe^{3+} \longrightarrow Hg^{2+} + Fe^{2+}$$

and excess ferric alum estimated by titration with 0.01 N ammonium thiocyanate. This method is capable of determining down to only 500 mg Hg kg$^{-1}$ of tissue sample.

**Spectrophotometric methods**

Various workers have described methods for the determination of mercury in tissues. Miller and Lillis[121] have described methods for the determination of phenylmercury acetate in urine, kidney, liver, muscle, spleen and brain. For urine, the sample (containing 5–20 μg of phenylmercury acetate) is refluxed with N sodium hydroxide, cooled, and excess 5% potassium permanganate solution added. Then 30% hydroxylamine sulfate: aqueous ammonia (1:1) and 30% ammonium sulfamate solution are added. The mixture is cooled and sufficient 12 N hydrochloric acid added under the surface of the liquid with vigorous swirling to lower the pH to not greater than 1. After further cooling, the solution is shaken well with purified chloroform (11 mL) for one min., the chloroform layer washed with 1 N hydrochloric acid and the washings rejected. The chloroform layer is separated, diluted with chloroform to 1 L mL$^{-1}$ and the extinction measured at 620 nm.

For kidney, liver, muscle or spleen, 1 g of the sample is treated as described above. Results obtained by these methods are accurate to within approximately ±1 μg of the amount of phenylmercury compound present.

Gage[122] has described a method for the trace determination of phenylmercury acetate in biological material. In this method an acidified aliquot of a 5% aqueous homogenisate of the tissue is extracted with benzene (20 mL). A 15 mL portion of the extract is shaken with 1% aqueous sodium sulfide solution and the evaporated aqueous layer is oxidised with potassium permanganate. To the oxidised solution, decolorised with hydroxyl-ammonium chloride solution, are added urea and EDTA; the pH is adjusted to 1.5. Chloroform is added and the solution is titrated with dithizone solution until the colour of the chloroform layer is intermediate between the orange of the mercury complex and the green of the dithizone solution. Recoveries of added phenylmercury salts and methylmercury salts from rat tissue and rat urine were low by up to 15% but concentrations down to 1 mg kg$^{-1}$ can be measured. Inorganic mercury does not interfere.

Ashley[123] has reviewed procedures for the determination of micro amounts of mercury in biological materials involving destruction of organic matter, use of dithizone for mercury extraction and the avoidance of simultaneous extraction of copper. Ashley[123] describes an absorptiometric method for determining the mercury–dithizone complex and also a photometric method in which the separated mercury is volatilised and its concentration in the vapour determined by means of a detector cell with a monochromic light source, the output of a photocell being measured and referred to calibration measurements.

Jones and Nickless[14] have also discussed the use of dithizone in the determination of methyl mercury salts.

**Atomic absorption spectrometry**

Various workers have discussed the application of this technique to the determination of organomercury and inorganic mercury in various biological materials[57] and in fish[4,27,47,124,125,81,131,108,133].

Methyl mercury compounds have been specifically dealt with by various workers[14,126,127,128,128]. Hanna and Tyson[18] determined total organomercury in urine by a flow injection atomic absorption spectrometric procedure involving on and off line oxidation of the organomercury species. Shum et al.[130] carry out a toluene extraction of fish, then treat the extract with dithizone and acidic cuprous chloride solution to form methyl mercury dithizonates which are then determined in amounts down to 0.08 mg kg$^{-1}$ by graphite furnace atomic absorption spectrometry.

The determination of organomercury in solid environmental samples such as fish requires low temperature preparation techniques to prevent losses of these volatile compounds. Thus, Uthe et al.[94,116] use a sulfuric and nitric acid digestion at 58°C to extract mercury from fish tissue. After digestion the extract is treated with an excess of 6% potassium permanganate solution and left for 18 hours. Precipitated manganese is then dissolved by addition of 30% hydrogen peroxide. Mercury is then released from the sample using a stannous sulfate hydroxylamine reduction system and estimated by flameless atomic absorption spectrometry. Application of this method to the determination of methyl mercuric chloride and mercuric chloride in fish samples showed that it was applicable to mercury levels down to 0.1 mg Hg $kg^{-1}$ fish with a standard deviation of 0.008 at the 0.1 mg $kg^{-1}$ level and 0.6 at the 9 mg $kg^{-1}$ level. Other workers, e.g. Society for Analytical Chemistry[127,134] and Agemian and Chau[81], have used mixtures of hydrogen peroxide and sulfuric acid for the determination of organic matter prior to the determination of mercury by cold vapour atomic absorption spectrometry.

Whilst the method discussed above is sensitive enough in some applications, in many types of biological samples it is necessary to determine much lower levels of organomercury. Thus, whilst the environmental level of mercury in canned tuna or human hair is about 0.5 mg $kg^{-1}$ the level in human urine is considerably lower (i.e. 0.5 ng $L^{-1}$). Oda and Ingle[39] have described a procedure for the determination of these levels of organomercury and inorganic mercury in tuna fish, hair and urine. They describe a speciation scheme for ultra trace levels of mercury in which inorganic and organomercury are selectively reduced by stannous chloride and sodium borohydride, respectively. The volatilised elemental mercury is determined by cold vapour atomic absorption spectroscopy. The samples were first digested by heating in 2 dram capped vials at 90°C for 15–30 mins, urine (1 mL) or hair (20–60 mg) or tuna (0.1–0.3 g) with 2.5 mL 10 M potassium hydroxide.

After digestion, the resultant solutions were centrifuged to separate remaining particulates and the supernatant liquid was decanted into a 100 mL volumetric flask. The vials were washed three to four times with 1% (w/v) concentrated nitric acid with centrifuging before each decant. When the sample had been transferred, 7.5 mL of concentrated nitric acid and 1 mL of 1% (w/v) potassium dichromate were added and the remainder of the volume was diluted with 1% (w/v) sodium chloride. These results are tabulated in Table 6.10 which shows that the requirement of a detection limit of 0.5 ng $g^{-1}$ for mercury in urine has been met.

Total concentration of mercury in hair (relative standard deviations of 6–11%) was fairly consistent with existing data which indicated mercury concentrations in hair ranging from 1 to 25 mg $kg^{-1}$ for samples from rural to industrial areas. About half of the mercury found was in the organic form concentrated by the body whilst the other half was probably in the form of externally absorbed $Hg^{2+}$.

Tuna samples showed about 0.5 mg $kg^{-1}$ total mercury contamination which is approximately normal for canned tuna. This is also the upper limit allowed

**Table 6.10.** Analysis of hair, urine and tuna. (From Oda and Ingle[39]. Reproduced by permission of American Chemical Society.)

| Sample | Inorganic | Amount of Hg(II) (ng g$^{-1}$) | |
|---|---|---|---|
|  |  | Organic | Total |
| Urine | 3.2 | 1.1 | 4.3 |
| Urine | 2.9 | 0.80 | 3.7 |
| Hair | $2.1 \times 10^3$ | $1.9 \times 10^3$ | $4.0 \times 10^3$ |
| Hair | $2.3 \times 10^3$ | $2.0 \times 10^3$ | $4.3 \times 10^3$ |
| Tuna | 35 | $4.1 \times 10^2$ | $4.5 \times 10^2$ |
| Tuna | 39 | $4.4 \times 10^2$ | $4.8 \times 10^2$ |

by the United States Food and Drug Administration. About 92% of the tuna appears as organomercury (probably mainly methylmercury). Relative precisions were 18 and 7% for inorganic and organomercury, respectively. The relative precision for the urine measurements was about 5 and 10% for inorganic and organomercury, respectively. The inorganic form predominates as expected since most organomercury that is introduced to the body is absorbed or broken down before excretion.

**Emission spectrometry**

Hanamura et al.[135] applied thermal vaporisation and plasma emission spectrometry to the determination of organomercury compounds and inorganic mercury in fish.

This method which operates at a plasma temperature of 5500 K and utilises the 253.65 nm mercury line has sufficient sensitivity to detect 0.1 µg mercury. The sample (0.25 mg) in a quartz crucible is heated over the range 25 to 450°C at 15°C min$^{-1}$ and as each organomercury compound is volatilised it passes into a 2540 M H$_2$ plasma torch operated at 500W and it is determined.

The analytical results of the emission signal of the analyte species are recorded as a function of the temperature of the samples, resulting in a plot similar to that obtained in differential thermal analysis. Because analyte species in the solid sample vaporise at different temperatures, each species produces a peak at a temperature characteristic of the analyte species and the sample type. The vaporised molecular species, introduced into the microwave induced plasma, appear as peaks at characteristic temperatures dependant primarily upon the molecular form and secondarily upon the sample composition.

**Inductively coupled plasma mass spectrometry**

Beauchemin[136] determined organomercury compounds in biological reference materials by inductively coupled plasma mass spectrometry using flow injection analysis. Methyl mercury was the only significant organomercury compound detected and it constituted at least 40% (lobster) and up to 90% (dogfish) of the total mercury present in the reference materials (dogfish muscle, lobster, hepatopancreas). Organomercury was extracted as the chloride with toluene and back extracted with aqueous cysteine acetate in chloride medium.

## Gas chromatography

This technique is essential if it is required to obtain an unequivocal identification of the type of organomercury compound occurring in a biological material as opposed to the total organic plus inorganic mercury content that is provided by atomic absorption spectrometry. An ideal combination is to use gas chromatography for separation of the organomercury compounds combined with a flameless atomic absorption or an inductively coupled plasma atomic absorption spectrometer as a detector system. Much of the original work on the application of gas chromatography to the identification and determination of organomercury compounds in biological materials were performed by Westhoo. In view of the comparatively high mercury content of fish found in Swedish lakes and rivers, Westhoo et al. embarked on an extensive survey of the nature and the concentration of mercury in fish from these waters (Westhoo[66,67,137]).

He describes a combined gas chromatographic and thin-layer chromatographic method (Westhoo[66,137]) for the identification and determination of methylmercury compounds in fish, in animal foodstuffs, egg yolk, meat and liver. He has also used a combination of gas chromatography and mass spectrometry to identify and determine methylmercury compounds in fish[138].

To extract organically bound mercury from muscle tissue of fish, Westhoo homogenised the fish with water and acidified with concentrated hydrochloric acid (1/5 of the volume of the suspension). Organomercuric compounds were then extracted in one step with benzene using the method described by Gage[122]. Methylmercury either originally present or added to the fish, could be extracted only with difficulty, when only a small amount of acid was added (e.g., at pH 1). From an aliquot of the benzene solution organomercury could be extracted with ammonium or sodium hydroxide solution, saturated with sodium sulphate for elimination of lipids. The yields were low and variable, but could be improved as described below.

Several workers have found that a clean-up procedure is necessary to remove fatty acid and amino acids, which could otherwise poison the gas chromatography column. The clean up is achieved by adding to the organic phase a reagent, such as sodium sulfide[122], cysteine[66,67,137], sodium thiosulfate or glutathione which forms a strong water soluble alkylmercury complex to extract the mercury complex into the aqueous phase. A halide is added to the aqueous phase, and the alkylmercury halides formed are re-extracted into an organic phase. Aliquots of this phase are finally injected into the gas chromatograph.

The mercury compound in the shellfish that caused the Minimata disease (Japan) was methyl(methylthio)mercury. Westhoo concluded that it is reasonable to assume that methylmercury, if present in Swedish fish, should at least to some extent be a methylthio derivative. The Hg-S bond is stronger than Hg-NH or Hg-OH bonds. Accordingly, it prevents the formation of these bonds, which should be produced by the ammonium hydroxide solution and increase the solubility in water. Any methylthio group present should therefore be removed before the extraction with alkali.

Distillation of the benzene extract at reduced pressure at room temperature or at 760 mm Hg pressure at 80°C to 1/10 of the original volume removed the factor that prevented an acceptable extraction by ammonium or sodium hydroxide solution (probably methanethiol and perhaps hydrogen sulfide). After the distillation and subsequent extraction with ammonium hydroxide solution the extract was acidified with hydrochloric acid and the organomercury compound was extracted once with benzene. After drying with anhydrous sodium sulphate, the benzene solution was ready for gas chromatography and, after concentration, also for thin-layer chromatography.

In the above procedure about 30% of the methylmercury was lost, mainly by unfavourable partition coefficients. In a model experiment of the benzene extraction of methylmercury from a hydrochloric acid solution, for instance 14% of the methylmercury was left in the water layer. The losses by partition are, however, characteristic of the compounds involved and reproducible. Consequently, they can be included in the calibration curve, thus disturbing the results only slightly. The yields can be increased by repeated extractions but good results are obtained with the above simple procedure. The calibration curve is based on the partition laws for methylmercury chloride, through some methylmercaptide and perhaps sulphide are probably present in fish. However, when hydrogen sulphide or methanethiol was added (30 µg per 5 µg mercury as methylmercury) to the aqueous phase before the first extraction, the 5 µg point was unaltered on the calibration curve. Large amounts of these sulphur compounds disturbed the analysis because they were not completely removed by the distillation.

When known amounts of methylmercury dicyandiamide were added to salt-water fish (frozen cod, Gadus morrhua, or haddock, Gadus aeglefinus), 82–95% of the additions were recovered.

Westhoo[66,67,137] used an electron capture detector and 150 cm × 3 mm (60in × $\frac{1}{8}$ in) stainless steel columns filled with Carbowax 1500 (10%) on Teflon 6, and washed DMCS. Nitrogen was used as carrier gas and column temperatures were 130–145°C. He identified methylmercury chloride in pike caught in the Baltic Ocean at concentrations between 0.07 and 4.4 mg kg$^{-1}$ of fish.

Westhoo[137] pointed out that if methylmercury attached itself to a sulfur atom by reaction with a thiol or hydrogen sulfide then the nonvolatile HgS compound produced would not be included in the determination. He has developed a modification to this method, to render it applicable to a wider range of foodstuffs (eggyolk and white, meat, liver, or fish), by binding interfering thiols in the benzene extract of the sample to mercuric ions added in excess or by extracting the benzene extract with aqueous cysteine to form the cysteine methylmercury complex.

Westhoo et al.[138] reported results obtained by gas chromatography with electron capture and with mass spectrometric detection on a range of samples of fish (Table 6.11). Total mercury was also determined on these samples by neutron activation analysis[139]. Results obtained by the three methods agree within ± 10% of the average value.

It was mentioned above that in the Westhoo method[137] for organomercury compounds in fish low recoveries are obtained unless the benzene extract of the fish homogenate is boiled to remove volatile mercaptans prior to extraction with ammonia. This distillation procedure was assumed to remove volatile thio compounds binding part of the methylmercury and preventing its uptake into ammonia.

When, however, small amounts of methylmercury dicyanidiamide (less than 0.05 mg kg$^{-1}$) are added to meat, liver or eggyolk and analysed according to the above method, the methylmercury was completely lost in liver and eggyolk, and only partly recovered from meat. After addition of 10 mg kg$^{-1}$ of methylmercury to meat or liver, most of it was recovered from meat, but only 5% from liver. Such a failure of the procedure can be expected; if the methylmercury in the neutralised extracts from these food-stuffs is firmly attached, exclusively or to a considerable, extent, to thiol groups of nonvolatile compounds, but only if the methylmercury salts formed are insoluble in alkali solutions. Model experiments showed, in fact, that after the addition of excess of methanethiol or thiophenol to methylmercury chloride in benzene, an extraction with 2N aqueous ammonia or with sodium hydroxide did not extract the mercury compound from the benzene layer.

Westhoo[67], examined problems associated with the determination of methylmercury salts in egg yolk and white with low methylmercury content, liver,

**Table 6.11.** Comparison between results for mercury levels in fish flesh, determined by combination gas chromatograph-mass spectrometer, gas chromatograph with electron capture detector, and activation analysis. (From Westhoo et al.[138] Reproduced by permission of Sveavägen, Stockholm.)

|  | Methylmercury (mg Hg kg$^{-1}$ fish flesh) | | Total Hg (mg kg$^{-1}$) |
|---|---|---|---|
|  | GLC-mass spectrometric measurement of $^{202}$Hg$^+$ | Gas chromatography with electron capture detector | fish flesh Neutron activation analysis |
| Pike 1 | 0.14 | 0.17 | Not determined |
| Pike 2 | 0.55 | 0.54 | 0.59 |
| Pike 3 | 2.53 | 2.57 | 2.70 |
| Pike 4 | 0.43 | 0.41 | 0.39 |
| Pike 5 | 0.49 | 5 | 0.54 |
| Pike 6 | 0.75 | 0.66 | 0.63 |
| Pike 7 | 0.72 | 0.70 | 0.66 |
| Pike 8 | 3.19 | 3.29 | 3.12 |

aquaria sediments and sludge, also bile, kidney, blood, meat and moss, in many of which mercury could not be accurately determined by the mercuric chloride method or the cysteine method. By combining these two methods, however, he was able to obtain good results with these various types of samples.

Excess mercuric ions were added to an aqueous liver suspension containing known amounts of a methylmercury salt. The analysis was performed according to the cysteine acetate modification. More than 100% of the methylmercury was recovered. When the acidified liver suspension containing mercuric ions was kept at room temperature overnight, the recovery increased, indicating a synthesis of methylmercury ions from mercuric ions by the liver under the conditions used.

For eggyolk with a low content of methylmercury the cysteine acetate procedure gave less than 90% recovery. With the combined method using cysteine and mercuric ions the recovery of methylmercury salt decreased almost to zero. But for sediments in aquaria and sludge, which similarly could not be analysed by the original cysteine acetate modification, the combined method gave good results.

A further attempt to improve the recovery in the cysteine acetate method involved a precipitation of the proteins in liver by molybdic acid. This increased the recovery of added methylmercury salt to about 90%. In eggyolk with a low content of methylmercury compounds, however, neither molybdic acid nor phosphomolybdic acid improved the results.

In a further attempt to overcome interference by organic sulfur, acidic sodium bromide has been used to extract organomercury as the bromide. Cuprous chloride was added to mask the sulfur compounds and displace any mercury bound to sulfur. In the presence of divalent sulfide ion, methylmercury compounds form bis (dimethylmercury) sulfide. This is insoluble in aqueous cysteine acetate which, in the Westhoo procedure, is added to extract the organomercury compounds from benzene solution in order to free them from interfering thiols. By the addition of cuprous chloride, the bis (dimethylmercury) compounds are converted into methylmercury chloride; this is extracted into benzene and then into aqueous glutathione. After acidification of the aqueous phase with hydrochloric acid, methylmercury chloride is re-extracted into benzene and determined by gas chromatography on a column (40 cm × 2 mm) packed with 25% of poly(diethylene glycol succinate) on Celite (60–80 mesh) or 10% poly(butanediol succinate) on Chromosrob W (60–80 mesh) with nitrogen as carrier gas and electron-capture detection.

Following the classic work on fish analysis of Westhoo, other workers[30,64,70,127,140–142] developed gas chromatographic methods for the analysis of biological materials including fish[70,143,145], urine[70], hair[70,144], sediments[70], seeds[143], grain[70], faces[70], milk[70], and tissue[142]. Callum et al.[142] have described the use of the proteolytic enzyme subtilisin Carlsberg Type A for the breakdown of human and animal tissue prior to the release of methylmercury. This enzyme has a high, non-specific proteolytic activity, which gives excellent breakdown of protein. The yields obtained are greater than those found using the more conventional acid hydrolysis. Cysteine hydrochloride and cupric bromide were also incorporated into the separation scheme. The benzene extract was analysed by gas chromatography with electron capture detection on a column comprising 5% ethylendigycol adipate on Gas Chrom W at 155°C.

In Table 6.12 is given a comparison of results obtained by this procedure compared with the values obtained by a method involving extraction with acid sodium bromide alone.

Ealy et al.[63] described a gas chromatographic method for the gas chromatographic determination of methyl, ethyl, and methoxyethyl mercury halides as their iodides in inorganic sediments, aquatic systems, seeds and fish.

Newsome[146] has described a method for the determination of methylmercury in fish and cereal grain products in which the sample (10 g) is homogenised for 10 min with N hydrobromic acid—2 N potassium bromide (60 mL) and filtered through glass wool. The combined filtrate is extracted twice with benzene. The combined benzene layers are extracted with a cysteine acetate solution, an aliquot of which is acidified with 48% hydrobromic acid and extracted with benzene. The benzene extracts are submitted to gas chromatography on a glass column (40 cm × 4 mm) packed with 2% of butanediol succinate on Chromosorb W (AW-DCMS) (100–120 mesh) operated at 120°C with nitrogen as carrier gas (80–100 mL min$^{-1}$) and $^3$H foil electron capture detector. The sensitivity of the method is in the range 0.01–0.90 mg Hg kg$^{-1}$. Mean recovery generally exceeds 95%. When direct gas chromatographic methods are used in the determination of alkylmercury compounds, interferences are often a problem, especially with the electron-capture detector, which is sensitive to other halogen compounds.

Longbottom[64] cooled the gases from the flame ionisation detector and led the gases through an atomic absorption spectrometer, but reported that it was less sensitive than the electron-capture detector for dialkyl mercury compounds. Bye and Paus[141] solved this problem by leading the effluent from the gas chromatographic column through a steel tube in a furnace at a temperature at which the organic mercury molecules are cracked. The products are then led through a 10 cm quartz cuvette placed in the beam from a hollow-cathode lamp in an atomic absorption spectrometer. These workers state that for many of the earlier methods, the calibration curves are obtained from measurements of peaks from pure standard solutions of organic mercury compounds. They doubt the

**Table 6.12.** Analysis of tuna and dry fish. (From Callum et al.[142] Reproduced by permission of Royal Society of Chemistry.)

|  | Method | Methylmercury (mg kg$^{-1}$) | Range | Coefficient of variation (%) |
|---|---|---|---|---|
| Tuna | Enzyme | 0.693 | 0.661–0.734 | 3.7 |
|  | Sodium bromide | 0.912 | 0.812–1.006 | 5.5 |
| Fish homogenate | Enzyme | 0.057 | 0.043–0.069 | 15.4 |
|  | Sodium bromide | 0.346 | 0.329–0.362 | 3.6 |

correctness of such a procedure, because it does not take into account the fact that appreciable amounts of mercury may be lost during the many extraction steps used in the analysis, especially in work with small samples and small volumes and state that a standard addition procedure should be used for calibration, and the standard organic mercury solution should be added as early as possible in the procedure.

A Perkin-Elmer model 800 gas chromatograph was used. The following operating conditions were satisfactory: column, 10% SP2300 on Chromosorb W 80–100 mesh; oven temperature 145°C; inlet temperature 200°C; carrier gas, nitrogen at a pressure of 3.5 kPa cm$^{-2}$ measured at the g.c. inlet; flow rate, 90 mL min$^{-1}$.

The Perkin-Elmer model 303 atomic absorption spectrometer was run at the 254 nm mercury line. Deuterium background correction was essential. 0.5 g portions of frozen fish were transferred to a tissue grinder, 0.5 µL of 1 M copper sulphate solution was added to each, and 50–100 µL of the standard mercury solution were added to two of the samples. Following a fairly detailed work-up procedure in which the sample is treated successively with bromine, sodium thiosulphate and potassium iodide a final benzene extract is obtained for gas chromatography.

Bye and Paus[141] detected methylmercury (not ethyl or phenylmercury) in fish samples. Ranges up to 10 mg kg$^{-1}$ Hg for methylmercury and ethylmercury chloride in mixtures were measured. Fish samples were found to contain 2.2 mg kg$^{-1}$ of mercury as methylmercury.

Uthe et al.[147] have described a solvent extraction gas chromatographic method for the determination of down to 0.01 mg kg$^{-1}$ alkylmercury compounds in crustacea. Kamps and McMahon[70] and workers at the Society for Analytical Chemistry[127] and Hight[148,149] and Holak[150] utilised gas chromatography with an electron capture detector for the determination of down to 0.25 mg kg$^{-1}$ of methylmercuric chloride in swordfish, shark, shrimp, oyster, clams and tuna fish.

Bache and Lisk[140] determined methylmercury compounds in fish by chromatography on a 60 cm glass column of Chromsorb 101 or 20% 1:1 OV-17/QF-1. Detection of the separated organomercury compounds was achieved by measurement of the emission spectrum of the 253.7 nm atomic mercury line which gave a linear response over the range 0.1–100 mg of injected methylmercury chloride. Average recoveries of methylmercury chloride in fish were 62% at the 0.3 mg kg$^{-1}$ level.

Cappon and Crispin Smith[70] have described a method for the extraction, clean-up and gas chromatographic determination of alkyl and arylmercury compounds and inorganic mercury in blood, grain, faeces, fish, crustacea, hair, milk, sediment, soft tissue and urine. Detection limits for fish and crustacea were, respectively, 0.02 and 0.001 mg kg$^{-1}$. Methyl, ethyl and phenylmercury are first extracted as the chloride derivatives. Inorganic mercury is then isolated as methylmercury upon reaction with tetramethyltin. The initial extracts are subjected to thiosulfate clean-up and the organomercury species are isolated as the bromide derivatives.

Total mercury recovery ranges between 75 and 90% for both forms of mercury, as assessed by using appropriate $^{203}$Hg labeled compounds for liquid scintillation spectrometric assay. Specific gas chromatographic conditions allow detection of mercury concentrations of 1 ng kg$^{-1}$. Mean deviation and relative accuracy average 3.2 and 2.2%, respectively. The accuracy and precision of this procedure was evaluated by analysing different sample types fortified with mercuric chloride and methylmercuric chloride. Results were cross checked by an atomic absorption procedure. Results obtained on samples by both methods are given in Table 6.13. There is good agreement between the two methods for samples containing methyl, ethyl and inorganic mercury and this is expressed in terms of gas chromatographic/atomic absorption ratios.

Panaro et al.[151] determined methyl mercury in fish by gas chromatography-direct current plasma atomic emission spectrometry.

Fischer et al.[152] determined methylmercury compounds in fish using gas chromatography with an atomic absorption detector and sodium methyl borate (NaB(C$_2$H$_5$)$_4$) derivatisation.

The sample was dissolved in methanolic potassium hydroxide then ethylated by derivatisation with sodium methyl borate. The reaction products were trapped cryogenically then analysed by gas chromatography. Down to 4 pg g$^{-1}$ (as Hg) of CH$_3$Hg$^+$ could be determined.

Decadt et al.[265] determined methylmercury in bird liver and kidney samples using a headspace injection system coupled to a gas chromatograph-microwave induced plasma system. Vapour concentrations decreased in the following order: CH$_3$HgI>CH$_3$HgBr>CH$_3$HgCl. The methylmercury compounds in the sample were transformed into methylmercury iodide by reaction with iodoacetic acid

$$(CH_3)_2Hg + ICH_2COOH = CH_3HgI + CH_3CH_2COOH$$

The detection limit in fish organ samples was 1.5 µg L$^{-1}$ of homogenate.

Table 6.13. GC-AA intercomparison study. (From Cappon and Crispin Smith[70]. Reproduced by permission of American Chemical Society.)

| Sample | GC | mg L$^{-1}$ Hg AA (as MeHg) | GC/AA |
|---|---|---|---|
| Fish | 1.10 | 1.06 | 1.04 |
| Hair | 266.2 | 272.9 | 0.98 |
| Muscle | 0.27 | 0.70 | 1.03 |
| | | as EtHg | |
| Blood | 0.72 | 0.77 | 0.94 |
| Kidney | 0.66 | 0.68 | 0.97 |
| | | as inorganic | |
| Blood | 0.59 | 0.57 | 1.47 |
| Fish | 0.08 | 0.07 | 1.14 |
| Sediment | 0.17 | 0.19 | 0.89 |

## Column chromatography and HPLC

Liquid chromatography using differential pulse electrochemical detection has been used to determine organomercury cations in tuna fish and shark meat[153]. The differential pulse mode of detection offers a substantial increase in selectivity over ampermometry.

Following alkaline hydrolysis the sample (1 g) is acidified with hydrochloric acid. The organomercury cations can then be extracted from the aqueous solution with toluene as the neutral chloride complexes. The aqueous back extraction solution used was 0.01 mol $L^{-1}$ disodium thiosulfate buffered to pH 5.5 with 0.05 mol $L^{-1}$ ammonium acetate. This extraction solution was compatible with the column chromatographic separation, and the determination was performed directly on this aqueous extract after filtering through a 0.2 μm syringe filter. In all cases, a standard addition procedure was used for the determination with known amounts of diluted $CH_3Hg^+$ solution added to the solid material before the hydrolysis step. The recovery was checked by comparison to a standard curve and found to be about 95%.

Various interference effects on the determination of organomercury compounds and how they are overcome are discussed by these workers.

Table 6.14 shows the results obtained when the method was applied to standard NBS fish samples. The sample chromatograms were characterised by a single response for methylmercury with high signal-to-noise ratio. Ethyl and phenylmercury were not detected in these samples.

The results obtained (see Table 6.14) for the methylmercury content of the fish samples were in fairly close agreement to the total mercury (as measured by alternate technique such as atomic absorption and neutron activation analysis).

Holak[150] used high performance liquid chromatography with an atomic absorption spectrometric detector to determine down to 0.004 μg (absolute) of methylmercury compounds in fish. MacCrehan and Durst[153] achieved a detection limit of 0.002 mg $kg^{-1}$ for methylmercury compounds in fish utilising HPLC with an electrochemical detector.

Table 6.14. Methylmercury content of fish samples. (From Mac Crehan and Durst[153]. Reproduced by permission of American Chemical Society.)

| Sample | Mercury species (ng $kg^{-1}$) | | | |
|---|---|---|---|---|
| | $MeHg^+$ | $EtHg^+$ | $PhHg^+$ | Total Hg |
| RM 50 Albacore Tuna | 0.93 ± 0.1 | nd[a] | nd | 0.95 ± 0.1 |
| Japanese shark paste | 8.41 ± 0.1 | nd | nd | 7.4 |

[a] nd = not detected.

## Substoichiometric analysis

Substoichiometric analysis is based on isotope dilution analysis. This method offers an accurate and precise determination of trace amounts of elements by measurement of radioactivity alone without corrections for chemical yield.

This technique has been applied to the determination of organomercury compounds in environmental samples. Kauda and Suzuki[154] have applied substiochiometric isotope dilution analysis to the determination of inorganic mercury and organically bound mercury in hair. Thionalide (thioglycolic-$\beta$-amino-naphthalide) was used as the extracting agent.

To determine total mercury the sample plus $^{203}$Hg was refluxed with sulphuric acid and 30% hydrogen peroxide and gently heated until pale yellow. Mercury was extracted with methyl isobutyl ketone. This phase was adjusted to pH 6.5 and was extracted with EDTA solution to recover mercury in an aqueous phase. The acidity of the aqueous solution was adjusted to 0.1 M in sulphuric acid and then the substoichiometric extraction of Hg(II) was carried out with 5 mL of 1 µM thionalide in chloroform. The organic phase was washed with 0.1 M sulphuric acid solution. To determine methylmercury the sample plus $^{203}$Hg methylmercury was treated with 3 M hydrochloric acid. The extract was shaken for about 5 min with two 10 mL portions of benzene to extract the methylmercury. Methylmercury in the combined benzene extract was back extracted into 5 mL of 0.02 M sodium sulphate solution at pH 6.5 and the substoichiometric extraction was carried out with 5 ml of 0.5 µm thionalide in chloroform. The radioactivity of the two extracts was measured as only a small proportion of the mercury is extracted in substoichiometry, the extractions can be carried out repeatedly from the same solution. From the amounts of radioactivity added to the original sample and that found in the final extracts and from the concentration of total mercury found in the final extract, it is possible to obtain the mercury content of the original samples. Down to approximately 1 µg of mercury can be determined by this procedure.

## Thin layer chromatography

Thin layer chromatography was carried out by Westhoo[137] either on the original methylmercury chloride containing fish extract or on derivatives prepared from this extract, such as dithizonate, bromide, iodide or cyanide. Light petroleum:diethylether (70:330) was used as developing solvent, using aluminium oxide or silica gel plates. Separated organomercury compounds were detected with saturated ethanolic solution of Michler's thioketone in ethanol.

Methylmercury dithizonate and phenylmercury dithizonate could be separated from each other in the fish extracts by thin-layer chromatography on aluminium oxide (limit of detection 0.2 µg). Methylmercury cyanide, chloride, bromide and iodide were separated by thin-layer chromatography on silica gel (limit of detection of the chloride and bromide: 0.02 µg).

## Miscellaneous

Miller and Wachter[155] have used a procedure based on reduction and digestion with sulfuric acid for the determination of low concentrations of mercury in biological materials.

Stuart[156] used $^{203}$Hg-labeled methyl mercuric chloride for *in vivo* labeling of fish to study the efficacy of various wet ashing procedures.

Pillay *et al.*[103] used neutron activation analysis to determine down to 0.01 mg kg$^{-1}$ of organomercury compounds in fish.

Robert and Robenstein[17] also used this technique in their indirect detection of Hg$^{199}$ NMR spectra of methylmercury complexes (e.g. CH$_3$Hg(II) thiol ligands) in environmental samples.

### 6.1.9 IN AIR

**Classical procedure**

Christie *et al.*[157] describe procedures capable of determining mercury at levels down to 10 mg m$^{-3}$ in air. In one method the air sample (500 mL) is passed at 50 L min$^{-1}$ through a column containing active carbon, and the carbon, freed from moisture by a stream of dry air, is then removed for determination of mercury. In a second method air (500 L) is drawn at 33.3 L min$^{-1}$ through a glass-fiber pad treated with cadmium acetate and sodium sulfide. The pad is then removed for determination of mercury. The active carbon or the cadmium sulfide pad is ignited. The colour produced on selenium sulfide test paper is compared with a colour chart. Both methods are applicable to the determination of ethylmercury (chloride and phosphate), diphenylmercury and methylmercury dicyandiamide. The first method only is applicable to diethylmercury.

Polley and Miller[99] have described a method for the determination in air samples of amounts of methyl-and ethylmercury chlorides down to 1–5 µg in 50–100 mL of sample. An alternate method developed by these workers (Miller and Polley[158]) is best used for the determination of amounts above 10 µg of methyl and ethylmercury chloride in 25 mL of the carbonate phosphate absorber solution previously mentioned by these workers (Kimura and Miller[159]). Large aqueous sample volumes are not deleterious in this method as they are in the direct method of analysis mentioned above. This is because of a favourable distribution coefficient of p-tolyl mercury chloride between the chloroform and water phases. The Miller and Swanberg[160] method is suitable for the determination of below 30 µg of alkylmercury compounds in sample sizes of up to 100 mL of carbonate-phosphate absorber solution.

**Atomic absorption spectrometry**

In earlier atomic absorption methods the air sample is passed through an absorber containing a liquid such as acidified potassium permanganate (Kimura and

Miller[3]) or a solid such as activated charcoal[157] silver or gold (Long et al.[161]), which collect not only elemental mercury but also alkyl or arylmercury halides and dialkyl or dialkyl mercury compounds. After desorption from one trap released mercury is determined by atomic absorption spectrometry at 257.3 nm. These methods do not have the required sensitivity for environmental air samples.

Dumareov et al.[167] have described a method which operates down to the 5 µg Hg m$^{-3}$ air level. In this method mercury is trapped on gold coated sand and the desorbed volatile mercury compounds determined by atomic absorption spectrophotometry.

Field samples were collected by drawing air through the absorber with a small pump (KNF, No. 5 ANE membrane pump) at flow rates of 2.5–3.5 L min$^{-1}$. A prefilter (Whatman GF/A, 13 mm diameter) is used to retain particulated and solid mercury compounds. The volume of air is measured by a calibrated dry gas meter. The volume sampled depends on the expected mercury concentrations. The loaded samplers can be stored for several days without change in mercury content. The mercury was desorbed by heating the absorber to 800°C and sweeping the products by a carrier gas stream into an atomic absorption spectrophotometer.

Schroeder and Jackson[163] determined mercury in the atmosphere using a method of selective preconcentration followed by pyrolysis and cold vapor atomic fluorescence detection to determine different mercury species collected from the atmosphere. Other studies have been performed by using sequential specific absorption tubes which separate different chemical forms of mercury by selective collection. In these later studies, mercury compounds were thermally desorbed and recollected on gold surfaces prior to elution into an emission detector. While these methods represented a significant advance in atmospheric mercury sampling by achieving the separation of volatile species of mercury, the analytical methods prevented positive identifications of the compounds by converting all forms to elemental mercury prior to detection. Chromatographic substrates have been used successfully for the collection of organics and of organic mercury[164]. By logical extension, a chromatographic method of analysis would permit the positive identification of organic mercury compounds by comparison of sample elution times/volumes with standard compounds.

The electrostatic accumulation furnace for electrothermal atomic spectrometry technique[165,166] has been successfully used for the precise, simple and fast determination of mercury in air samples. It has also been shown that the electrostatic accumulation furnace can be easily operated with a collection efficiency of approximately 100% with particulate matter as well as with mercury vapour.

## Gas chromatography

Ballantine and Zoller[167] developed a mercury collection method that is compatible with a chromatographic method of analysis and is capable of detecting levels of organic mercury in the atmosphere as low as 0.1 ng m$^{-3}$.

Table 6.15. Analytical system parameters for the analysis of MMC, DMM, and total volatile mercury. (From Ballantine and Zoller[167]. Reproduced by permission of American Chemical Society.)

| Parameter | System | | |
|---|---|---|---|
| | Methyl mercury chloride | Dimethyl mercury | Total volatile Hg |
| Flow rate, mL/min | $85 \pm 5$ | $75 \pm 5$ | $60 \pm 5$ |
| Oven temp, °C | $165 \pm 5$ | same | same |
| Column (1 m × 6 mm o.d.) | 5% FFAP on Gas Chrom Q (80/100 mesh) | Chromosorb 101 (60/80 mesh) | empty glass column |
| Collection tube (TI) | Chromosorb 101 (60–80) (2–4 cm) Tenax GC (60/80 mesh) (5 cm) | Chromosorb 101 (5 cm cm) | gold coated glass beads |
| Description temp °C | 200 Detector | 90 | 350 |
| Microwave power | 26 W (forward), 1 W (refected) | slit width carrier gas pressure at outlet | 110 um argon 7.55 = 5 mmHg |
| Monochromator setting | 253.7 nm | | |
| Quartz capillary | 1 mm i.d. 6 mm o.d. | | |

Total volatile mercury air samples were passed through a column of gold coated glass heads. The mercury on the glass heads is directly eluted by heating at 500°C on to a column packed with Chromosorb 101 maintained at a temperature of 20°C to collect methylmercury chloride and 80°C to collect dimethylmercury. Collection efficiencies on these columns under these conditions are about 95%. The organomercury compounds were then released from the columns by heating, respectively, to 200 and 90°C and analysed under the conditions given in Table 6.15 detection being achieved by a microwave plasma detector operated at 257.3 nm.

Major disadvantages of this technique are the necessity of long collection times (3–12 h) when studying background levels, and relatively long analyses (approximately 1 h) for methylmercuric samples. The significant advantages over previous methods are the simplification of sample elution directly into the gas chromatographic column which minimises the possibilities of contamination and the ability to positively identify two of the most important organic forms of mercury present in the atmosphere.

### 6.1.10 PRESERVATION OF MERCURY CONTAINING SAMPLES

The problems of preserving mercury in solution are well known. Although controversy still exists over which preservative is the best, agreement on several of the factors which affect the stability of mercury solutions seems to have been reached. For example, it is agreed that low pH values, high ionic strengths and oxidising environments help in keeping mercury in solution. Acids such as sulfuric acid[168] nitric acid[169–175] and hydrochloric acid[176] have been widely used in different amounts. Oxidants such as potassium permanganate[177–184,42,58] and potassium dichromate[58,182,6,42] have been shown to prevent volatilisation of mercury. Sodium chloride and gold[182,42] have also been used as preservatives. Various workers have commented on the instability of mercury solutions when stored in polyethylene or polypropylene containers[175,185–187].

McFarland[186] concluded that polythylene phials can be used for the neutron irradiation of mercury solutions without loss. Weiss and Chew[175] carried out neutron irradiation of aqueous and nitric acid solutions of mercury in polyethylene containers and showed that, whilst no losses occurred in the presence of nitric acid, both absorption and volatilisation losses of up to 18% occurred in aqueous medium. Heiden and Aikens[185] studied the effect of differences in commercial polyethylene bottles on the stability of parts per billion mercury(II) solutions.

Coyne and Collins[187] showed that solutions containing 0.05–0.5 mg mercury in the presence of acetic acid and formaldehyde preservative lost about 50% of mercury in 3 days. Loss of mercury appeared to be related to the original concentration of mercury and the presence or absence of a preservative. Nitric acid (added to give a final pH of 1.0) was moderately effective in preventing loss of mercury provided that the acid was placed in the container before the sample was added.

Other workers have compared the effectiveness of glass with various plastic containers for the safe storage of dilute mercury solutions[58,169].

Rosain and Wai[169] concluded that mercury losses in solutions which were pH 7 with respect to nitric acid were more rapid from distilled water than from natural water and were most severe from poly(vinyl chloride) containers. These workers recommended that sampling for mercury determination be done in glass or polyethylene containers, that the sample be acidified to a pH of less than 0.5 with nitric acid, and that analysis be carried out as soon as possible after sampling.

Feldman[58] stored 0.1–10 mg inorganic mercury solutions in glass and polyethylene containers in the presence of various reagents. He found that the solutions lost appreciable amounts of mercury even when solutions of 1–5% nitric acid, 0.5% sulphuric acid–0.01% potassium permanganate or nitric acid–0.01% dichromate were added. However, solutions could be stored in glass for up to 5 months in the presence of 5% nitric acid–0.01% dichromate (added as potassium chromate or dichromate) and in polyethylene for at least 10 days in the presence of 5% nitric acid–0.05% dichromate.

Carron and Agemian[188] have pointed out that, whilst the majority of fresh water samples rarely contain mercury at levels over 0.5 $\mu$g L$^{-1}$ and in most cases 0.2 $\mu$g L$^{-1}$, most previous investigators of the stability of mercury solutions have carried out their tests at higher mercury levels. These workers studied preservation methods which provide both low pH values as well as oxidising environments using both synthetic and natural samples in a variety of containers, in order to obtain a practical method which would be adaptable to routine analysis for mercury in natural waters at sub µg L$^{-1}$ levels by the automated cold vapour atomic absorption technique. The essential requirements was that the preservation method should maintain mercury in waters of low salt content (low conductivity such as distilled water) and of high salt content (high conductivity). The outcome of this work was that Carron and Agemian[188] recommended glass containers washed with concentrated nitric or chromic acid and a preservative consisting of a mixture of 1% sulfuric acid and 0.05% potassium dichromate. This preservatives gives good accuracy, precision, and low detection limits. It was also observed that the presence of methylmercury ions improves preservation efficiency.

Hawley and Ingle[41] studied the breakdown of a 1.0 µg L$^{-1}$ methylmercury chloride solution caused by 1.0% (v/v) nitric acid alone, 0.01% (w/v) potassium dichromate alone and a mixture of 1.0% (v/v) nitric acid and 0.1% (w/v) potassium dichromate which are used as preservatives for total mercury. Measurements of inorganic and methylmercury content were made within hours of preparation and after 1,3 and 8 days of standing in 100 mL glass volumetric flasks at room temperature. The results were compared to those obtained with an unpreserved 1 µg L$^{-1}$ methylmercury chloride solution and are shown in Figure 6.3. About 20% of the methylmercury was observed to be converted to inorganic mercury (the form easily reducible by stannous chloride) under these conditions in slightly over a day (Figure 6.3). The total amount of mercury (inorganic and organic) in solution remained fairly constant over a 3 day period with an approximate 25 ±8% loss over a period of 8 days. Comparison of decomposition induced by 0.01% potassium dichromate alone (Figure 6.4) and 1.0% nitric acid (Figure 6.5) alone to that caused by the combination of the two reagents indicates that the major factor appears to be the presence of nitric acid. Nitric acid alone converts almost half of the methylmercury to mercuric ions in just about 3 days and losses in terms of total mercury from the solution amount to about 26 ±5% over 8 days. The potassium dichromate is not nearly as destructive although about 15% of the CH$_3$Hg$^+$ is decomposed in 3 days while maintaining greater than 90% effectiveness in retaining total mercury for more than a week.

It should be noted that, in preparing any of the test solutions, the acid or dichromate was diluted to 50–75 mL with water before addition of methylmercury chloride solution so that the organomercury compound was never in direct contact with the concentrated preservation reagent. The unpreserved methylmercury (Figure 6.6) retained its concentration remarkably well over a 3 day period and as expected a minimum of methylmercury breakdown was observed over that time. However losses of about 33 ± 12% of the total mercury concentration after 8 days was noted, again as expected since no preservatives were present. The presence of inorganic mercury at the end of the test period may be partially attributed to photon-induced decomposition since methylmercury is somewhat sensitive to light.

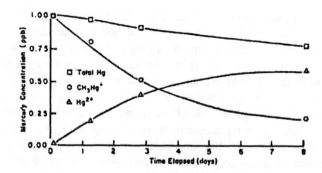

**Figure 6.3.** Preservation study of 1 μg L$^{-1}$ CH$_3$HgCl with 1.0% HNO$_3$ and 0.01% K$_2$Cr$_2$O$_7$. (From Hawley and Ingle[41]. Reproduced by permission of American Chemical Society.)

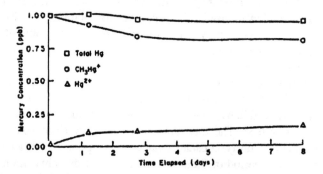

**Figure 6.4.** Preservation study of 1 μg L$^{-1}$ CH$_3$HgCl with 0.01% K$_2$Cr$_2$O$_7$. (From Hawley and Ingle[41]. Reproduced by permission of American Chemical Society.)

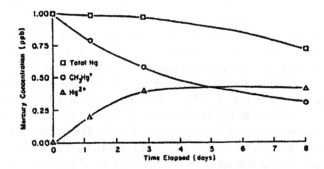

**Figure 6.5.** Preservation study of 1 μg L$^{-1}$ CH$_3$HgCl with 1.0% HNO$_3$ alone. (From Hawley and Ingle[41]. Reproduced by permission of American Chemical Society.)

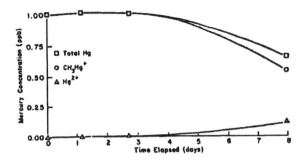

**Figure 6.6.** Preservation study of 1 µg L$^{-1}$ CH$_3$HgCl with no preserving reagents. (From Hawley and Ingle[41]. Reproduced by permission of American Chemical Society.)

This study indicates that, where speciation of mercury is the primary objective, the use of nitric acid should be avoided to minimise decomposition, or if added, the analysis must be run as soon as possible, preferably within hours of the addition. However, with extended periods of preservation, acid and dichromate should be used and a determination of total mercury can be obtained with a fair amount of accuracy since losses would be minimised. However, the original speciation information for the sample is no longer determinable unless analysis is carried out immediately after preservation.

## 6.2 ORGANOLEAD COMPOUNDS

### 6.2.1 NATURAL AND POTABLE WATERS

**Atomic absorption spectrometry**

Various workers have described methods for the determination of organolead compounds in river and natural waters[191−197]. Chau et al.[198] extracted dialkyl and trialkyllead compounds into benzene from aqueous solution. Following their chelation with dithiocarbamate, they are then converted to their butyl derivatives for gas chromatography atomic absorption spectrometry. A detection limit of 0.1 µg lead L$^{-1}$ was achieved with one litre of water.

De Jonghe et al.[199] have developed a method for the determination in water of traces of triethyllead compounds without interference from mono-, di- and tetraalkyllead compounds and inorganic lead.

After enrichment of the sample by a fast vacuum distillation technique and saturation of the residual volume with sodium chloride, the analytes are extracted in chloroform. By incorporation of specific purification steps, interference from other forms of organic and inorganic lead is completely eliminated. The final chloroform extract is treated with sulfuric acid solution in order to transfer the trialkyllead compounds present back into an aqueous solution. The analysis is completed by graphite furnace atomic absorption spectrometry. A detection limit of 0.02 μg can be achieved with 1 L samples. In contrast to earlier approaches where other forms of organic and inorganic lead may give rise to serious interferences, the determination is highly specific for trialkyllead compounds, even in the presence of up to 100 μg $L^{-1}$ of inorganic lead salts. For environmental applications a 1 L volume is advised. After shaking for 1 min with hexane (1 mL per 100 mL of water) the sample is filtered on a Type RA Milliprefilter of 1.2 μm pore size. The water is brought into a rotary evaporator and evaporated under vacuum at 60°C until a residual volume of ca. 15 mL remains. The sample is then quantitatively transferred to a 100 mL separating funnel and the volume is adjusted to 25 mL with distilled water. After this, the sample is extracted twice with 25 mL portions of chloroform. The extracts are combined and 1 mL of a 1 mM dithizone in chloroform solution is added. Next, this organic phase is shaken for about 2 min with an aqueous solution consisting of 15 mL of an ammonium citrate/ammonia buffer and 35 mL of 0.1 M EDTA. The chloroform layer is separated off, extracted first with distilled water to remove traces of EDTA and buffer and then for 1 min with 5 mL of 0.1 N sulfuric acid solution. Analysis of this sulfuric acid extract by means of graphite furnace atomic absorption spectrometry allows the quantitation of the trialkyllead initially present.

The overall accuracy obtained in this procedure was 87 ± 4% for trimethyllead and 92 ± 5% for triethyllead.

It was found that in this method up to 1000 μg of inorganic lead and up to 100 μg of dialkyllead and tetraalkyllead compounds (in the sample portion taken) can be analysed without exceeding the limit of detection.

**Gas chromatography**

Several methods for the determination of tetraalkyllead compounds in water have been proposed which depend on a combination of gas chromatography with selective detectors. There have been few reports on the direct gas chromatographic determination of trialkyllead salts[193] and none for dialkyllead salts. Forsythe and Marshall[194] approach to the determination of these salts was to further alkylate them with a Grignard reagent to convert them to their tetraalkyl analogues prior to capillary column gas chromatography using electron capture detection. Ammoniacal buffer (pH 8.5 10 mL) was added to the sample which was then extracted three times with 0.5 mg $L^{-1}$ dithizone in 50% benzene/hexane. The organic extracts were combined, reduced in volume to 0.5 mL and derivatised directly by reaction with phenyl magnesium bromide in tetrahydrofuran medium. Following the addition of water, organolead compounds were extracted with a small volume of hexane.

A feature of the gas chromatographic apparatus to avoid sample decomposition was the provision of an allglass insert and modified injector, which increased internal volume and decreased metal surface area.

Forsythe and Marshall[194] used a 30 m fused silica DB-1 column as this gave a superior separation of alkylleads from coextractives. Helium was used as carrier gas and nitrogen was used as make-up gas relative to 5% argon methane because it resulted in more stable detector operation. Nitrogen doped with 10 µL L$^{-1}$ oxygen caused increased detector response time.

A linear increase in detector response was observed with increasing analyte concentration (range 4–500 pg for EtMe$_2$PbPh, Et$_2$MePbPh and EtMePbPh$_2$). Recoveries of dialkyl and trialkyllead compounds obtained from water by this procedure was consistently high in the 0–20 µg L$^{-1}$ lead range.

Gas chromatography/mass spectrometry identified the following mixed alkylphenylleads, EtMe$_2$PbPh, Et$_2$MePbPh and EtMePbPh$_2$ as well as biphenyl and terphenyls in the crude re-equilibration reaction mixtures. Forsythe and Marshall[194] were not able to detect any of those 'mixed' alkylleads during recovery trials of alkylphenyllead standards or during recovery trials using trialkyllead chlorides or dialkyllead chlorides.

Chakraborti et al.[200] determined ionic alkyllead compounds in water using a combination of gas chromatography with atomic absorption detection. Analysis of 500 mL samples enabled the determination of 1.25 ng L$^{-1}$ for PbMe$_3$$^+$ and 2.5 ng L$^{-1}$ for PbEt$_2$$^+$. Extraction recoveries were in excess of 90%.

Rapsomankis et al.[192] studied the speciation of lead and methyllead ions in water using gas chromatography with atomic absorption detection. Methyllead compounds were first ethylated using sodium tetraethylborate. Using the purge and trap technique with an atomic absorption detector a detection limit of 0.2 pg L$^1$ was achieved for PbMe$_3$$^+$ and PbMe$_2$$^{2+}$, when 50 mL water samples were used.

Various workers[197,201,202,193,203] have discussed the application of gas chromatography combined with an atomic absorption detector for the determination of organolead compounds in water.

Chau et al.[197,201,202] have described a simple and rapid extraction procedure to extract the five tetraalkyllead compounds (Me$_4$Pb, Me$_3$EtPb, MeEt$_2$Pb, MeEt$_3$Pb and Et$_4$Pb) in hexane extracts of water samples. The extracted compounds are analysed in their authentic forms by a gas chromatographic–atomic absorption spectrometry system. Other forms of inorganic and organic lead do not interfere. The detection limit for water (200 mL) is 0.50 µg L$^{-1}$. An average recovery of 89% was obtained by this procedure for the aforementioned alkyllead compounds.

The main interest of Chau et al.[201] was in the determination of organically bound lead produced by biological methylation of inorganic and organic lead compounds in the aquatic environment by microorganisms. The gas chromatographic-atomic absorption system used by Chau et al.[201] (used without a sample injection trap) for this procedure has been described[197]. The extract was injected directly into the column injection port of the chromatograph. Instrumental parameters were identical. A Perkin Elmer electrodeless discharge lead lamp was used; peak areas were integrated.

In this method 2000 mL water sample and 5 mL of high purity hexane were placed in a 250 mL separatory funnel and shaken vigorously for 30 min in a reciprocating shaker. The solution was allowed to stand for about 20 min for phase separation. Approximately 195 mL of the water was drained off and the remaining mixture transferred into a 25 mL tube with a Teflon lined cap. Without separating the phases, a suitable aliquot (5–10 µL) of the hexane, was injected into the gas chromatograph–atomic absorption spectrometric system.

To calibrate the method a known amount (5 µg) of standard tetramethyllead, was added to the hexane layer after injection of the sample, mixed gently, and centrifuged again if necessary. The same volume as used in sample analysis was injected. The increase in peak area due to the standard added was used to calculate the amount of tetraalkyllead in the sample. It is not necessary to separate the phases or to know the volume of hexane after extraction.

The calibration curves for each of the five tetraalkyllead compounds expressed as lead were identical and linear up to at least 200 ng above which overlapping peaks occurred. If only one compound was present (e.g. tetramethyllead) the curve was linear up to at least 2000 ng. Although ionic forms of lead such as diethyllead dichloride and trimethyllead acetate do not extract in the benzene phase, any lead compounds that distribute into the benzene phase as tetraalkyllead will be determined. Chau et al.[201] found that environmental samples can contain other forms of organolead compounds that are extractable into benzene but which are not volatile enought to be analysed by the gas chromatographic–atomic absorption spectroscopic technique, hence the need for a speciation specific analytical system.

In Table 6.16 are tabulated recoveries obtained by the procedure for five tetraalkyllead compounds form water samples.

The application of a combination of gas chromatography and atomic absorption spectrometry to the determination of tetraalkyllead compounds has also been studied by Chau et al.[202] and by Segar[203]. In these methods the gas chromatography flame combination showed a detection limit of about 0.1 µg lead. Chau et al.[202] have applied the silica furnace in the atomic absorption unit and have also shown that the sensitivity limit for the detection of lead can be enhanced by three orders of magnitude.

Table 6.16. Recovery of tetraalkyllead compounds from water. (From Chau et al.[201] Reproduced by permission of American Chemical Society.)

| Compound | Water | | |
|---|---|---|---|
| | added (µg) | found (µg) | recovery (%) |
| $Me_4Pb$ | 10.00 | 8.78 | 87.8 ± 3 |
| $Me_3EtPb$ | 13.15 | 11.80 | 89.7 ± 4 |
| $Me_2Et_2Pb$ | 14.30 | 12.50 | 87.4 ± 3 |
| $MeEt_3Pb$ | 10.15 | 9.08 | 89.5 ± 4 |
| $Et_4Pb$ | 14.20 | 12.82 | 90.3 ± 7 |
| Average | | | 88.9 ± 7 |

*Four determinations for each sample.

The system used by these workers consisted of a Microtek 220 gas chromatography and a Perkin Elmer 403 atomic absorption spectrometer. These instruments were connected by means of stainless steel tubing (2 mm o.d.) connected from the column outlet of the gas chromatograph to the silica furnace of the atomic absorption spectrometer. A four-way valve was installed between the carrier gas inlet and the column injection port so that a sample trap could be mounted, and the sample could be swept into the g.c. column by the carrier gas. The recorder (10 mV) was equipped with an electronic integrator to measure the peak areas, and was simultaneously actuated with the sample introduction so that the retention time of each component could be used for the identification of peaks.

The furnace was constructed from silica tubing (7 mm diameter, 6 cm long) with open ends. The lead compounds separated by gas chromatography were introduced to the centre of the furnace through a side-arm. Hydrogen gas was introduced at the same point at a flow rate of 1.35 mL min$^{-1}$; the burning of hydrogen improved the sensitivity. The furnace was wound with 26 gauge Chromel wire to give a resistance of about 5 $\Omega$. The voltage applied to the furnace was about 20 V a.c. regulated by a variable transformer so that the furnace temperature with the hydrogen burning was about 1000°C. The silica furnace was mounted on top of the a.a.s. burner and aligned to the light path.

The sample trap was a glass U-tube (6 mm diameter, 26 cm long) packed with 3% OV-1 on Chromosorb W, which was immersed in a dry ice–methanol bath at ca. −70°C as described by Chau et al.[202] A known amount of gaseous sample was drawn through the trap by a peristaltic pump operated at 130–150 mL min$^{-1}$. After sampling, the trap was mounted to the four-way value and heated to ca. 80–100°C by a beaker of hot water, and the adsorbed compounds were swept into the gas chromatographic column. Liquid samples can be directly injected into the column through the injection port, without a sample trap.

**Gas chromatography**

Analyses were carried out on a Pye Model 104 gas liquid chromatograph coupled to a Pye Unicam electron capture detector operated in the pulsed mode at 300°C with a 63 Ni source. Samples were injected onto a 150 cm glass column packed with Chromosorb W (60–85 mesh) coated with 10% SE-30. Column temperature and flow rate of nitrogen carrier gas were adjusted to give the optimum retention for the tetraalkyllead present: samples suspected of containing tetraethyllead were chromatographed at 100°C with a flow rate of 60 mL min$^{-1}$; those suspected of containing tetramethyllead, at 64°C and 40 mL min$^{-1}$. Confirmation of the identity of the suspected tetraalklyllead was obtained by a rechromatographing with a genuine sample and for all except tetramethyllead, by rechromatographing the sample on a 300 cm glass column packed with a 10% 1,2,3-tris(2-cyanoethoxy) propane on Chromosorb W at 64°C and 40 mL min$^{-1}$ flow rate. Peak areas were integrated electronically and quantified by comparison with those of standards of tetraalkyllead in benzene.

## Thin-layer chromatography

For higher concentrations of tetraalkyllead, its identity was confirmed by thin-layer chromatography. A sample of water was extracted into 40:60 petroleum ether (50 mL) and iodine monochloride (0.3 mL) was added to the separated petroleum ether extract. The solution was shaken, allowed to stand for 10 min and extracted into water (10 mL). The water from the aqueous extract was evaporated under reduced pressure and the residue taken up in acetone. After evaporation of the bulk of the acetone, the sample was run by thin-layer chromatography against samples of diethyllead, dichloride and dimethyllead dichloride. Tetraalkyllead added to tetraalkyllead-free samples showed the presence of $Pb^{2+}$ ions by thin-layer chromatography at concentrations down to 2 mg $L^{-1}$ in the petroleum ether extract.

## Analysis for alkyllead salts

Sample (10 mL) of the aqueous extract suspected of containing alkyllead salts were analysed by a colorimetric technique using pyridylazoresorcinol, the absorption of the resulting solution being measured at 515 mm on a spectrophotometer. Confirmation of the presence of dialkyllead salts was attempted by evaporation of the water from the aqueous extract at reduced pressure, dissolution of the residue in acetone, and analysis by thin-layer chromatography. For trialkyllead salts the aqueous extract was saturated with sodium chloride and extracted into benzene; the benzene extract was then concentrated and analysed by thin-layer chromatography. These techniques also identified the alkyl groups present, but the sensitivity was restricted by the limits of detection of the thin-layer chromatographic analysis.

Chau et al.[198] developed a gas chromatographic method for the determination of ionic organolead species and applied it to samples of Lake Ontario water. When one litre of sample water is taken the following species can be determined in amounts down to 0.1 μg $L^{-1}$: $Me_2Pb^{2+}$ $Me_3Pb^+$, $Et_2Pbt^{2+}$ $Et_3Pb^+$ and $Pb^{2+}$.

The highly polar ions are quantitatively extracted into benzene from aqueous solution after chelation with dithiocarbamate. The lead species are butylated by a Grignard reagent to the tetraalkyl form, $R_nPbBu_{(4-n)}$ (R=$CH_3$ or $C_2H_5$) and $Bu_4Pb$, all of which can be quantified by a gas chromatography–atomic absorption spectrometry method. Molecular covalent tetraalkyllead species, if present in the sample, are also extracted and quantified simultaneously. Other metals co-extracted by the chelating agent do not interfere.

In this procedure, to a 1 L sample of lake water is added 50 mL of aqueous 0.5 M sodium diethyldithiocarbamate, 50 g of sodium chloride and 50 mL of benzene, and the mixture is shaken for 30 min. The benzene phase is then carefully evaporated in a rotary evaporator to 4.5 mL in a 10 mL centrifuge tube to which 0.5 mL of butyl Grignard reagent is added. The mixture is gently shaken for 10 min, and washed with 5 mL of 0.5 M sulfuric acid to destroy the excess of Grignard reagent. About 2–3 mL of the organic phase is pipetted into a small vial and dried with anhydrous sodium sulfate. Appropriate amounts (5–10 µL) are injected into the gas chromatography–atomic absorption system. The chromatographic column was of glass, 1.8 m long, 6 mm diameter, packed with 10% OV-1 on Chromosorb W (80–100 mesh) with nitrogen carrier gas flow rate of 65 mL min$^{-1}$. The injection port and transfer line were at 150 and 160°C respectively; the column was programmed from 80°C to 200°C at a rate of 5°C min$^{-1}$. The 217.0 nm line from a lead electrodeless discharge lamp operated at 10 W was used with an electrically heated quartz furnace at 900°C with hydrogen flowing at 85 mL min$^{-1}$ and a deuterium lamp was used for background correction. Peak areas were integrated with an Autolab Minigrator (Spectra-Physics, CA).

All the ionic alkyllead compounds slowly degrade in the presence of light. However, lake water samples enriched with dimethyllead chloride and trimethyllead acetate at 100 µg L$^{-1}$ level are stable over a period of at least one month in the laboratory when stored in the dark and refrigerated. There is no need to add any preservative to the sample. Storage in a cold dark room is recommended. Alternatively, the samples can be extracted, butylated and dried over anhydrous sodium sulfate.

Extraction efficiencies for various ionic alkyllead species from water by this procedure are tabulated in Table 6.17.

Tetraalkyllead compounds (typically tetramethyllead, trimethyllead, dimethyl diethyllead, methyltriethyllead and tetraethyllead) have high vapour pressures and are seldom found in water unless they are adsorbed on sediments or particulate matter. If these compounds are present in the water sample, they can be extracted into benzene and included in the determination. The butylation reaction has no effect on the tetraalkyllead. A chromatogram showing the separation of five tetraalkylleads, four ionic alkyllead and $Pb^{2+}$ species in a synthetic sample is illustrated in Figure 6.7.

Acidic water (pH 1–2) lowered the extraction efficiency to about 10%. Therefore samples should be adjusted to pH 6–7 before extraction. Many metals are extracted by sodium diethyldithiocarbamate but those which extracted and

Table 6.17. Extraction efficiency of dialkyllead, trialkyllead, and lead (II) ions species from water[a]. (From Chau et al.[198] Reproduced by permission of Elsevier Science Publishers BV, Netherlands.)

| Medium | Extraction efficiency (%)[b] | | | | |
| --- | --- | --- | --- | --- | --- |
| | $Me_2Pb^{2+}$ | $Me_3Pb^+$ | $Et_2Pb^{2+}$ | $Et_3Pb^+$ | Pb(II) |
| NaCl (sat.) | 10(3) | 40(4) | 14(4) | 95(3) | 0 |
| NaCl (sat.) + KI (40 g) | 45(3) | 100(5) | 58(6) | 112(8) | 5(4) |
| Tropolone | 17(4) | 25(4) | 20(7) | 15(7) | 20(6) |
| NaDDTC | 109(8) | 97(6) | 105(9) | 94(5) | 94(8) |
| NaDDTC + NaCl (sat.) | 98(5) | 100(7) | 97(6) | 98(6) | 93(5) |

[a] Distilled water, 100 mL lead compounds, 20 µg each species; volume of benzene in all cases, 5 mL; tropolone (0.5% in benzene), 5 mL; NaDDTC (0.5 M), 5 mL.
[b] Average of two results with average deviation in parentheses.

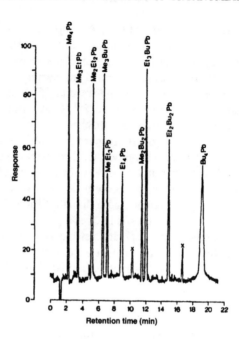

**Figure 6.7.** Gas chromatography–atomic absorption spectrometry of five tetralkyllead compounds (10 ng each); four butyl derivatives of dialkyl and trialkyllead (8 ng each) and $Pb^{2+}$ (15 ng); ×, unidentified lead compounds. (From Chau et al.[198]. Reproduced by permission of Elsevier Science Publishers, Netherlands.)

likewise butylated to the tetraalkyl metal forms are tin IV and germanium IV. When tetramethyltin and tetramethyl germanium were tested, there was no response whatsoever in the atomic absorption spectrometric detection system, set at 217.0 nm.

Estes et al.[193] described a method for the measurement of triethyl- and trimethyllead chloride in tap water, using fused silica capillary column gas chromatography with microwave excited helium plasma lead specific detection. Element specific detection verified the elution of lead species, a definite advantage to the packed column method. The method involved the initial extraction of trialkyllead ions from water into benzene, which was then vacuum reduced to further concentrate the compounds. Direct injection of the vacuum concentrated solutions into the gas chromatography–microwave excited helium plasma system gave detectability of triethyllead chloride at the 30 mg $L^{-1}$ level and trimethyllead chloride at the mg $L^{-1}$ level, but the method was time consuming and only semiquantitative.

The equipment used by Estes et al.[193] featured a gas chromatograph interfaced with a microwave induced and sustained atmospheric pressure helium plasma (9C-MEF) for element selective and sensitive detection. It incorporates a chemically deactivated, low-volume, valveless fluidic logic gas switching interface, to vent large quantities of eluent solvent which would disrupt the helium discharge as sustained by the $TM_{010}$ cylindrical response cavity. The neatness and venting characteristics of the interface are outstanding. The detection system features a low resolution scanning monochromator with a quartz retractor place background corrector directly after the entrance slit to improve selectivity ratios of elements whose emission wavelengths occur in the cyanogen background region. Use of the plasma can be used as a nonselective universal organic compound detector.

The gas chromatographic-microwave excited helium plasma detection system operating parameters are listed below:

GC-MED operating parameters for lead and carbon

| Parameters | For lead |
|---|---|
| column | |
| packing material | sp-2100 WCOT fused silica Carbowax pretreated |
| dimensions | 12.5 m, ×300 μm o.d. × 200 μm i.d. |
| injection split | 100 to 1 |
| carrier gas flow rate (helium) | 1 mL min$^{-1}$ |
| temperatures | |
| column or program | 140–185°C |
| injector | 180°C |
| transfer block | 180°C |
| interface oven | 180°C |
| total plasma flow rate | 125 mL min$^{-1}$ |
| PMT tube and voltage | RCA 1P28 700 V. |
| entrance and exit slit | |
| widths | 25 μm |
| height | 12 mm |
| microwave input power | 54 W |
| wavelength | 405.8 nm |
| picoammeter time constant | 0.10s |

In order to avoid thermal decomposition of trialkyllead compounds at active sites on the apparatus the quartz injection port liner, quartz discharge tube and the quartz interface tubing were removed and silanised by passing 100% dichlorodimethylsilane through the tubing. The tubing was placed in an airtight container flushed with nitrogen and allowed to react for 30 min. The tubes were washed with a large quantity of 'spectrograde' methanol in order to quench any remaining chlorosilane bonds and dried at 250°C for 1 h while being flushed with helium. This was followed by helium flushing at room temperature overnight. It is quite evident from examination of the gas chromatograms that the differences in triethyllead chloride response from the 'at plasma' and 'venting' column positions was negligible. Thus, the loss of triethyllead chloride when utilising the nonsilanised quartz interface tubing was due to the chemically active quartz surface.

Estes et al.[204] and Chau et al.[205] have also reported the $n$-butyl Grignard derivatisation of the trialkyllead ions extracted into benzene as the chlorides from spiked tap water which has been saturated with sodium chloride. A precolumn trap enrichment technique is substituted to replace solvent extract vacuum reduction. Final measurement of the lead compounds, now as $n$-butyltrialkylleads, is undertaken with the gas chromatograph-microwave emission detector system. Precolumn Tenax trap enrichment of the derived trialkylbutylleads enables determination to low μg $L^{-1}$ levels to be carried out. In this procedure the water sample (100 mL) is adjusted to pH 7, saturated with sodium chloride and extracted with a small volume of Specpure benzene.

A 5 mL portion of the benzene extract was placed in a centrifuge test tube and 1 mL of 2.0 M $n$-butyl-magnesium bromide Grignard reagent in tetrahydrofuran was added. The test tube was tightly stoppered and the solution was mixed thoroughly and was allowed to stand for 1 h. After $n$-butylation, 250 μL of a tetra-$n$-butyllead in benzene-internal-reference solution (0.316 mg $mL^{-1}$) was added. Excess Grignard remained in the mixture which could be stored at 0°C or lower without decomposition for at least 1 day if the subsequent analysis could not be performed immediately. A 95 ± 8% derivativization $n$-butylation recovery was obtained.

The direct quantitative gas chromatographic measurement of trimethyllead chloride or triethyllead chloride suffers from two major difficulties. (a) Both compounds are thermally unstable and tend to decompose even at the lowest possible injection port temperatures (ca. 160–170°C) required to give complete and rapid volatilisation, and (b) both compounds are very chemically reactive giving some tailing of chromatographic peaks even with the most inert chromatographic column available. Thus, direct quantitative gas chromatographic measurement of trialkyllead compounds is difficult.

If the trialkyllead compounds can be converted to tetraakylleads, quantitative determination is feasible. The use of $n$-butylmagnesium bromide Grignard reaction for trimethyl- and triethyllead chlorides appeared to be promising. Tetra-$n$-butyllead can be used as an internal reference which will not interfere with speciation of methyl or ethyl tetra- or trialkyllead compounds and should mark the termination of the lead specific analysis, since it should be the last tetraalkyllead compound to elute.

Over the spiked tap water concentration range investigated the extraction efficiency of trimethyllead chloride was 5.7 ± 0.6% and of triethyllead chloride was 93 ± 12%, a not unexpected result in view of the more ionic nature of the former compound. Detection limits were 35.0 μg $L^{-1}$ based on a 2 × noise signal for triethyllead chloride and 5.6 μg $L^{-1}$ for trimethyllead chloride. In addition, the detectability of both compounds could be improved by the use of a larger precolumn Tenax trap.

Other organolead compounds elute at detectable levels after the compounds of interest but before the internal reference. These compounds could result from decomposition impurities in the trimethyl- or triethyllead chloride, thermal redistribution products occurring from reaction in the injection port, impurities in the internal reference, or organoleads in the tap water. A tap water blank was carried through the analysis procedure and to ensure detection of organoleads other than the internal reference, three 30 µL samples were placed on the same trap before desorption. The additional organolead compounds were not seen in these experiments. Injection port thermal redistribution of alkyllead compounds is unlikely to occur; hence it seems likely that the unidentified lead peaks are derivatised species perhaps of decomposition impurities (i.e. di- and monoethyl and methylleads in the trialkyllead chloride standards).

This procedure offers several advantages for the simultaneous determination of trimethyllead and triethyllead ions in aqueous media. It is rapid and reproducible; it does not require the most strictly deactivated fused silica columns, since the chemically active trialkyllead chlorides are quantitatively converted to relatively chemically inert $n$-butyltrialkylleads. Thus, other capillary columns can be utilised with little effect on the quantitation, because the tetraalkylleads are so amenable to gas chromatography. The precolumn trap enrichment procedure eliminates the need to destroy excess Grignard reagent (to prevent column degradation) and the need for solvent extract vacuum reduction. Elimination of the vacuum reduction step solves the problems of trialkyllead chloride decomposition on the walls of the vacuum vessel and the loss of analyte compound due to volatilisation at reduced pressure. The use of an $n$-butyl Grignard reagent allows the simultaneous speciation of all organoleads of the methyl and ethyl alkyls (i.e., tetraalkyltrialkyl, dialkyl and monoalkyl) which might be extracted and derivatised.

### Polarography

Direct polarography has been used[206] to determine trialkyllead compounds in water[206]. Colombini et al.[207] have described a technique for the determination of organometallic species in natural waters based on selective organic phase extraction coupled with differential pulse polarography. The analytical procedure was applied to alkyl lead compound speciation and found to be reliable for the individual detection and determination of organolead complexes at trace levels in natural waters including sea water.

### Thin-layer chromatography

Potter et al.[208] have applied gas chromatography and thin-layer chromatography to the detection and determination of alkyllead compounds and alkyllead salts in natural waters. Samples suspected of containing alkyllead salts were applied from an appropriate solution to plates coated with 0.25 mm layer of MN Aluminoxid

G (Camlab, Cambridge) equilibrated with the atmosphere. Chloroform was found to be a suitable solvent for $R_3PbX$ and water or acetone for $R_2PbX_2$. Plates were eluted with acetic acid–toluene (1:19, v/v) and the spots were developed by spraying with a solution of dithizone in chloroform (0.1% w/v). $R_3PbX$ gave a yellow spot and $R_2PbX_2$ gave a salmon red spot; inorganic lead gave a crimson spot on the baseline and tetraalkyllead was not detected. Under these conditions the $R_f$ values for triethyllead chloride and trimethyllead chloride were 0.5 and 0.2, respectively and for diethyllead dichloride and dimethyllead dichloride were 0.3 and 0.1 respectively. The mixed methyllead salts gave distinct spots with intermediate $R_f$ values. The limit of detection of this method was 0.5–1.0 µ g alkyllead salt.

The total concentration of lead in natural waters is generally low owing to absorption onto sediment unless the pH is exceptionally low. This explains the observation that the levels of lead in rain water are twice those in ground water. Tetraalkyllead is insoluble in, and denser than, water and would be expected to accumulate with the sediment. Alkyllead salts are generally much more soluble in water than the corresponding inorganic lead salts and would remain in solution in the absence of suspended solids. However, alkyllead salts at low concentrations are totally absorbed onto a variety of natural sediments which are nearly always present in natural waters.

**Miscellaneous**

Jarvie et al.[209] studied the reactions between trimethyllead chloride and sulfides in aqueous systems. Factors affecting these reactions have been investigated. It is concluded that the formation of tetramethyllead from trimethyllead compounds in natural waters is due to chemical reactions, and not to biomethylation.

### 6.2.2 SEAWATER

Bond et al.[210] examined the interferences occurring in the stripping voltametric determination of trimethyllead in sea water by polarography and mercury-199 and lead-207 nuclear magnetic resource spectrometry. NMR and electrochemical data show that Hg(II) reacts with $(CH_3)_3Pb^+$ in seawater. Consequently, anodic stripping voltametric methods for determining $(CH_3)_3Pb)^+$ and inorganic Pb(II) may be unreliable. Solvent extraction followed by spectrophotometry have been used to determine trialkyllead compounds in water[330].

### 6.2.3 RAINWATER AND SNOW

**Gas chromatography**

Lobinski et al.[211] carried out speciation analysis of organolead compounds in Greenland snow at the femtogram per gram level by capillary gas chromatography using an atomic absorption detector. In this procedure the snow sample was mixed with EDTA buffered to pH 8.5 and extracted with hexane to preconcentrate organolead compounds. The extract was propylated using propyl magnesium chloride and the product analysed by capillary gas chromatography.

$Et_3Pb^+$ and $Et_2Pb^{2+}$ could be detected in amounts down to 0.02 and 0.02–0.5 µg kg$^{-1}$ respectively. Neither $Me_3Pb^+$ or $Me_2Pb^{2+}$ were found in the snow samples analysed.

**Column chromatography**

To preconcentrate trialkyllead species in rainwater Blaszkewicz et al.[212] complexed interfering metal ions in rain water with EDTA before adjustment of the pH to 10. Samples were pumped through an extraction column of silica gel to adsorb lead compounds which were then desorbed with acetate buffer containing methanol at pH 3.7. The eluate was diluted and adjusted to pH 8 with borate buffer before further concentration on a Nucleosil 10-C18 pre-column. Adsorbed trialkyllead compounds were eluted by backflushing on to a RP-C18 column and separated with methanolic acetate buffer. Online detection used a post-column chemical reaction detector. Detection limits for sample volumes of 500 mL were 15 µg L$^{-1}$ and 20 µg L$^{-1}$ for trimethyl- and triethyl-lead, respectively, Standard deviation was less than 4% for a sample containing 90 pg triethyllead per mL.

6.2.4 INDUSTRIAL EFFLUENTS AND WASTEWATERS

**Spectrophotometric methods**

Imura et al.[213] have described a spectrophotometric procedure employing 1-hydroxy-4-(4-nitrophenylazo)-2-naphthoate as a chromogenic reagent for the determination of triethyllead ions in industrial effluents. The coloured adduct is extracted with chloroform. The absorption maximum is at 440 nm and the optimum pH for the extraction from 1% aqueous sodium chloride is 8.1–8.3. In this determination of about 60 µg of triethyllead ions, dimethyl and diethyllead ions (100 µg) and $Pb^{2+}$ (1.8 mg) these components are masked with 0.01 M ethylenediamine NN'-bis (2-hydroxyphenylacetic acid) (disodium salt) and copper and ferrous iron are masked with 0.01 M 1,2-diaminocyclohexane-NNN'N'-tetracetate. Several other ions do not interfere. The limit of determination is 0.2 mg L$^{-1}$ of triethyllead.

**Atomic absorption spectrometry**

Aneva[214] used graphite furnace atomic absorption spectrometry to determine 4–100 µg L$^{-1}$ of tetraalkyllead compounds in waste waters. Tetraalkyllead compounds were extracted from the water with hexane, converted into water soluble iodides by reaction with iodine in the extract and re-extracted into dilute nitric acid. The mean recovery was 95% and no interference was experienced from other metallic or nonmetallic ions.

## 6.2.5 SEDIMENTS

Chau et al.[205] have described the optimum conditions for extraction of alkyllead compounds from sediments originating in nonsaline waters and in saline waters[201]. Analyses of some environmental samples revealed for the first time the occurrence of dialkyl- and trialkyllead in sediments in areas of lead contamination.

The various alkyllead species and lead(II) are isolated quantitatively by chelation extraction with sodium diethyldithiocarbamate, followed by $n$-butylation to their corresponding tetraalkyl forms, $R_n PbBu_{(4-n)}$, and $Bu_4 Pb$, respectively (R=Me, Et) all of which can be determined by a gas chromatograph using an atomic absorption detector. The method determines simultaneously the following species in one sample: tetraalkyllead ($Me_4Pb$, $Me_3EtPb$, $Me_2Et_2Pb$, $MeEt_3Pb$, $Et_4Pb$); ionic alkyllead ($Me_2Pb^{2+}$, $Et_2Pb^{2+}$; $Me_3Pb^+$, $Et_3Pb^+$); $Pb^{2+}$. Detection limits expressed for Pb were 15 µg kg$^{-1}$ for sediment samples.

In this method, the sediment (1–2 g) sample was extracted for 2 h in a capped vial with 3 mL of benzene after addition of 10 mL of water, 6 g of sodium chloride, 1 g of potassium iodide, 2 g of sodium benzoate, 3 mL of sodium diethyldithiocarbamate and 2 g of coarse glass beads (20–40 mesh). After centrifugation of the mixture, a measured aliquot (1 mL) of the benzene was butylated using 0.2 mL $n$-butyl magnesium chloride with occasional mixing for 10 m. The mixture was washed with 2 mL sulfuric acid (1 N) to destroy excess Grignard reagent. The organic layer was separated in a capped vial and dried with anhydrous sodium sulphate. Suitable aliquots were injected into the gas chromatograph.

Chau et al.[205] found that in spiking experiments on sediments both the diethyl and triethyl species were recovered at satisfactory levels (Table 6.18).

**Table 6.18.** Recovery and reproducibility of alkyllead and lead(II) compounds from sediment. (From Chau et al.[205] Reproduced by permission of American Chemical Society.)

| Amount of Pb added (µg) | Recovery[b] (%) | | | | |
|---|---|---|---|---|---|
| | Me$_3$Pb | Et$_3$Pb | Me$_2$Pb | Et$_2$Pb | Pb(II) |
| 1 | 113(9) | 73(14) | 103(6) | 104(15) | |
| 5 | 111(3) | 86(2) | 116(2) | 94(5) | |
| 10 | 122(4) | 106(1) | 114(3) | 85(1) | |
| 20 | 99(1) | 111(4) | 118(2) | 89(3) | |
| av | 111 | 94 | 113 | 93 | |
| % rel std dev | 4 | 4 | 14 | 15 | 9[c] |
| ($n = 6$) at 5 mg kg$^{-1}$ level | | | | | |

[a]Sediment 1 g; spiked compounds expressed as Pb.
[b]Average of two results with average deviation in parentheses.
[c]The sediment contained 71 mg kg$^{-1}$ of Pb(II) which was used to evaluate the reproducibility. No Pb(II) was added to sample.

Chau et al.[201] has described a simple and rapid extraction procedure to extract the five tetraalkyllead compounds ($Me_4Pb$, $Me_3EtPb$, $Me_2Et_2Pb$, $MeEt_3Pb$, $Et_4Pb$) from sediment. The extracted compounds are analysed in their authentic forms by a gas chromatographic-atomic absorption spectrometry system. Other forms of inorganic and organic lead do not interfere. The detection limits for sediment (5 g) is 0.01 mg kg$^{-1}$. In this method 5 g wet sediment and 5 mL of EDTA reagent, (0.1 M, 37 g $Na_2EDTA$ $2H_2O$ $^{-1}$), and 5 mL of hexane are placed in a 25 mL test tube with a Teflon lined screw cap and shaken for 2 h then centrifuged. The extract was analysed by gas chromatography using an atomic absorption spectrometer set at the 217 µm lead line and with a silica furnace as a detector[197]. This combination has a detection limit of about 1 µg lead, about three orders of magnitude better than can be achieved using a flame ionisation detector.

The system used by these workers consisted of a Microtek 220 gas chromatograph and a Perkin-Elmer 403 atomic absorption spectrophotometer. These instruments were connected by means of a stainless steel tubing (2 mm o.d.) connected from the column outlet of the gas chromatograph to the silica furnace of the atomic absorption spectrometer. The silica furnace was set at 1000°C. The gas chromatographic column was packed with 3% OV-1 supported on Chromosorb W. The column was temperature programmed at 15°C h to 150°C.

The furnace was constructed from silica tubing (7 mm i.d., 6 cm long) with open ends. The lead compounds separated by gas chromatography were introduced to the centre of the furnace through a side-arm. Hydrogen gas was introduced at the same point at a flow rate of 1.35 mL min$^{-1}$, the burning of the hydrogen improved the sensitivity. The silica furnace was mounted on top of the atomic absorption spectrometer burner and aligned to the light path. When the absorbances were plotted against lead concentrations, each of the five tetraalkyl compounds gave similar calibration curves; the response was linear up to at least 200 ng Pb, above which over-lapping of the peaks occurred. Reisinger et al.[315] also used gas chromatography with an atomic absorption spectrometric detector to determine organolead compounds in non saline sediments.

### 6.2.6 BIOLOGICAL MATERIALS

**Organolead in biological materials**

In the determination of these ionic alkyllead compounds in biological tissues, difficulties are further compounded by their strong affinity with protein and lipid matrices. There has been a dearth of information on the occurrence of these compounds in biological samples mainly because of the lack of suitable methodology. Up to the present time there have been only relatively few methods[202,193,198,205,216,217] dealing with the ionic alkyllead compounds in biological samples.

Several analytical methods for the determination of trialkyllead compounds have been reported: (a) the separation of triethyllead ion as the benzoate from liver and identification by infrared spectrometry; (b) the separation of trialkyllead compounds from rat blood, urine, brain, liver and kidney via a laborious multiple extraction separation procedure with final dithizone completion of the decomposed organoleads and colorimetric determination of the lead dithizone complex[193,217].

## Atomic absorption spectrometry

Chau et al.[202] applied gas chromatography atomic absorption to the determination of tetraalkyllead compounds in fish samples in high lead areas. Of some 50 fish samples analysed, only one sample was found to contain detectable amounts (0.26 mg kg$^{-1}$) of tetramethyllead in the fillet. Since there is no known tetraalkyllead industry and tetramethyllead is not used in gasoline in this area, the source of tetramethyllead is not yet known. The possibility that it comes from *in vivo* lead methylation in the sediment or in the fish cannot be totally disregarded.

Sirota and Uthe[216] have described a fast, sensitive atomic absorption procedure for determining tetraalkyllead compounds in biological materials such as fish tissue. Tissue homogenates were extracted by shaking with a benzene/aqueous EDTA solution, a measured portion of the benzene was removed and after digestion, the residue was defatted if necessary. The resultant $Pb^{2+}$ was determined by flameless atomic absorption spectroscopy using a heated graphite atomiser. Using a sample weight of 5 g, 10 μg kg$^{-1}$ of lead as $PbR_4$ can be determined with a relative standard deviation of 5%. No other forms of lead that were tested (e.g., $PbR_3X$, $PbR_2X_2$) were found to partition into the benzene layer under these conditions.

The recovery and selectivity of the method was evaluated by adding known amounts of different lead compounds to previously analysed tissue samples. The results obtained are summarised in Table 6.19 and indicate a satisfactory recovery and selectivity for tetraalkyllead compounds. Various marine tissues were sampled for total lead and tetraalkyllead. Results are summarised in Table 6.20. Di- and tri-substituted alkyl-leads were also evaluated in this system and the results were satisfactory.

## Gas chromatography

Chau et al.[201,205] have described a gas chromatographic method for the determination of the nanogram level of trialkyllead ions as the chloride in biological materials.

Chau et al.[201,202] have described a simple and rapid method for the determination of down to 0.025 μg kg$^{-1}$ $Me_4Pb$, $Me_3EtPb$, $Me_2Et_2Pb$, $MeEt_3Pb$ and $Et_4Pb$ in fish samples. This procedure, discussed in more detail in the section on water analysis above, involves homogenisation of the sample with EDTA and extraction with a small volume of hexane prior to analysis by gas chromatography utilising an atomic absorption detector. Using this method it was demonstrated that trout after exposure to water containing 3.5 μg L$^{-1}$ tetramethyllead concentrated this

Table 6.19. Recovery of tetraalkyllead compounds from cod liver homogenate. (From Sirota and Uthe[216]. Reproduced by permission of American Chemical Society.)

| Compound | Amount added ($\mu$g Pb) | Amount added (ng kg$^{-1}$) | Total Pb present prior to spike ($\mu$g) | Total Pb found after spike ($\mu$g) | Amount of spike found ($\mu$g) | Recovery (%) |
|---|---|---|---|---|---|---|
| Tetramethyllead | 0.10 | 20 | 0.25 | 0.38 | 0.13 | 130 |
| | 0.10 | 20 | 0.27 | 0.40 | 0.13 | 130 |
| | 0.50 | 100 | 0.06 | 0.575 | 0.515 | 103 |
| | 0.50 | 100 | 0.06 | 0.625 | 0.565 | 113 |
| Tetraethyllead | 0.10 | 20 | 0.14 | 0.21 | 0.07 | 70 |
| | 0.10 | 20 | 0.16 | 0.26 | 0.10 | 100 |
| | 0.50 | 100 | 1.056 | 1.548 | 0.492 | 98 |
| | 0.50 | 100 | 0.053 | 0.65 | 0.542 | 119 |
| | 0.50 | 100 | 0.045 | 0.42 | 0.375 | 75 |

Table 6.20. Concentrations of total lead and tetraalkyllead in various marine tissues. (From Sirota and Uthe[216]. Reproduced by permission of American Chemical Society.)

| Tissue | Concentration total Pb(mg kg$^{-1}$) | Concentration PbR$_4$ (mg kg$^{-1}$) | % Tetraalkyllead of total lead |
|---|---|---|---|
| Frozen cod (liver homogenate) | 0.39 ± 0.04 | 0.037 ± 0.003 | 9.5 |
| | | 0.010 ± 0.001 | |
| Large, freshly killed cod (liver homogenate) | 0.52 ± 0.05 | 0.125 ± 0.005 | 24 |
| Small, freshly killed cod, (2 separate lobes analysed). | A. 0.21 ± 0.04$^a$ | 0.028 | 13.3 |
| | B. | 0.044 | 20.9 |
| Lobster digestive gland (homogenate) | 0.20 ± 0.02 | 0.162 ± 0.004 | 81 |
| Frozen mackerel muscle (homogenate) | 0.14 ± 0.02 | 0.054 ± 0.005 | 38.6 |
| Flounder meal | 5.34 ± 1.02 | 4.79 ± 0.32 | 89.7 |

$^a$For total lead determinations both lobes.

substance in its tissues, mainly in lipid layers with concentration factors between 124 and 934.

Chau et al.[205] have also described a method for the determination of dialkyllead and trialkyllead compounds in fish. This method involves use of a tissue solubiliser to digest the sample followed by chelation extraction with sodium diethyldithiocarbamate, followed by $n$-butylation using butyl magnesium chloride to their corresponding tetraalkyl forms, $R_n$Pb Bu$_{(4-n)}$ and R$_4$Pb, respectively (R = methyl and ethyl). The method determines simultaneously in one sample; tetraalkyllead, ionic alkyllead (R$_2$Pb$^{2+}$ and R$_3$Pb$^+$), and divalent inorganic lead, all of which are determined by gas chromatography using an atomic absorption detector.

In this method, the fish samples were homogenised a minimum of five times. About 2 g of the homogenised paste was digested in 5 mL of tetramethylammonium hydroxide solution in a water bath at 60°C for 1–2 h until the tissue had completely dissolved to a pale yellow solution. After cooling, the solution was neutralised with 50% hydrochloric acid to pH 6–8. The mixture was extracted with 3 mL of benzene for 2 h in a mechanical shaker after addition of 2 g sodium chloride and 3 mL of sodium diethyldithiocarbamate. After centrifugation of the mixture, a measured amount (1 mL) of the benzene was transferred to a glass-stoppered vial and butylated with 0.2 mL of butyl magnesium chloride with occasional mixing for ca. 10 min. The mixture was washed with dilute sulphuric acid to destroy the excess Grignard reagent. The organic layer was separated in a cupped vial and dried with anhydrous sodium sulphate. Suitable aliquots (10–20 µL) were injected to the gas chromatographic atomic absorption system for analysis.

The recoveries of trialkyllead and dialkyllead species at different levels obtained by this procedure are shown in Table 6.21. The relative low recovery of dimethyllead is in agreement with the results of other investigators, Chau et al.[205] noticed that there was a large Pb(II) peak in the fish sample containing spiked dimethyllead, but such was not found in the standard which was run in parallel but without the sample. They attributed such low recovery to the decomposition of dialkyllead in the fish matrix. Diethyllead, however, did not decompose significantly and was recovered at near quantitative levels. For the first time, the occurrence of triethyl and diethyllead compounds was detected in fish samples and in other environmental materials (Table 6.22).

Chau et al.[205] have also applied derivatisation with butylmagnesium halides, followed by gas chromatography to the determination of mono, di, tri and tetraalkyllead compounds in biological samples. Detection was achieved by an atomic absorption detector.

**Table 6.21.** Recovery and reproducibility of alkyllead and lead (II) compounds from fish[a]. (From Chau et al.[205] Reproduced by permission of American Chemical Society.)

| Amount of Pb added (µg) | Recovery[b](%) | | | | |
|---|---|---|---|---|---|
| | Me$_3$Pb | Et$_3$Pb | Me$_2$Pb | Et$_2$Pb | Pb(II) |
| 1 | 72 (5) | 102 (5) | 79 (4) | 93 (0) | |
| 5 | 88 (4) | 88 (3) | 89 (5) | 103 (2) | |
| 10 | 93 (2) | 88 (2) | 56 (10) | 92 (2) | |
| 20 | 91 (2) | 81 (2) | 62 (6) | 114 (2) | |
| av | 86 | 92 | 71 | 101 | |
| % rel std dev | 15 | 7 | 18 | 20 | 14[c] |
| ($n = 6$) a$^+$ 5 mg kg$^{+1}$ level | | | | | |

[a]Fillet, 2 g; spiked compounds expressed as Pb.
[b]Average of two results with average deviation in parentheses.
[c]The fish fillet contained 142 mg kg$^{-1}$ of Pb(II) which was used to evaluate the reproducibility. No. Pb(II) was added to sample.

**Table 6.22.** Analysis of environmental samples (St. Lawrence River Near Maitland Ontario)[a]. (From Chau et al.[205] Reproduced by permission of American Chemical Society.)

| Sample | Me$_4$Pb | Me$_3$EtPb | Me$_2$Et$_2$Pb | MeEt$_3$Pb | Et$_4$Pb | Me$_3$Pb$^+$ | Me$_2$Pb$^{2+}$ | Et$_3$Pb$^+$ | Et$_2$Pb$^{2+}$ | Pb$^{2+}$ |
|---|---|---|---|---|---|---|---|---|---|---|
| Carp | 137 | —[b] | — | — | 780 | 2735 | 362 | 906 | 707 | 1282 |
| Pike | — | — | 96 | 142 | 7475 | 162 | — | 1215 | 1310 | 4133 |
|  | — | — | — | 169 | 1018 | 215 | — | — | — | 1040 |
|  | — | — | — | 146 | 1125 | 205 | — | 53 | — | 1187 |
| White sucker | — | — | — | — | 4384 | 196 | — | 3433 | 4268 | 3477 |
|  | — | — | — | 293 | 2948 | 95 | — | 2171 | 2196 | 3610 |
| Small mouth bass | — | — | 57 | 187 | 1204 | — | — | 223 | 92 | 254 |
|  | — | — | 71 | 252 | 1834 | — | — | 660 | 275 | 305 |
| Sediment | — | — | — | 142 | 1152 | — | — | 187 | 22 | 10000 |
|  | — | — | — | — | 309 | — | — | — | — | 5582 |
| Macrophytes, mixed surface | — | — | — | — | 68 | — | — | 132 | — | 4327 |
| 4 m deep | — | 38 | 1501 | 3613 | 16515 | — | — | 558 | 113 | 59282 |

[a]Data expressed in µg kg$^{-1}$ as Pb, wet weight; whole fish for fish samples.
[b]Not detectable.

The capillary column gas chromatographic method developed by Forsythe and Marshall[194] for the determination of di and trialkyllead compounds in water, described earlier, has also been applied to the determination of these compounds in whole egg samples. Alkyllead salts ($R_3Pb^+$ and $R_2Pb^{2+}$, R=Me or Et) are recovered from water or whole eggs by complexometric extraction with dithizone. The dithizonates are phenylated and speciated by capillary column gas chromatography on a 30 m fused silica DB-1 column with electron capture detection. The method is sensitive to low µg kg$^{-1}$ levels of lead salts in 2.5 g egg homogenate. At these levels methyllead salts (but not ethyllead salts) interact strongly with the sample matrix. Treatment of the matrix with lipids and proteases releases them.

**Whole egg hydrolysis procedure**

Whole egg homogenate was incubated at 37°C for 24 h in 60 mL of 5% ethanol/0.1 M phosphate buffer (pH 7.5) containing 30 mg of Lipase Type (III) and 30 mg of Protease Type (XIV) This technique was found to be 72 ± 9% effective after 24 h relative to classical acid hydrolysis.

Absolute ethanol (15–22 mL) and ammoniacal buffer (pH 9.5 10 mL) were added to the sample. The mixture was extracted three times with 0.5 g L$^{-1}$ dithizone (10 mL) in 50% benzene/hexane. The organic extracts were combined, centrifuged and back extracted three times with 10 mL of nitric acid (9.15 M). The aqueous washes were combined, neutralised with sodium hydroxide and further basified with 5 mL ammoniacal buffer (pH 9.5). The alkyl lead salts were recovered by extracting the aqueous phase three times with 0.1 g L$^{-1}$ dithizone (10 mL) in 50% benzene/hexane. These washes were combined and centrifuged and the organic layer was reduced in volume to 0.5 mL.

**Differential pulse anodic scanning voltammetry**

Birie and Hodges[218] have described a differential pulse anodic scanning voltammetry method for the determination of down to 0.01 mg kg$^{-1}$ of ionic lead compounds ($EtPb^+$, $MePb^+$, $EtPb^{2+}$ and $MePb^{2+}$) in oyster macuma.

## 6.2.7 PLANT MATERIALS

Chau et al.[205] have described a procedure involving gas chromatography with an atomic absorption detector for the determination of organolead compounds ($Me_3EtPb$, $Me_2Et_2Pb$, $MeEt_3Pb$, $Et_4Pb$, $Et_3Pb^+$ and $Et_2Pb^{2+}$) in macrophytes.

## 6.2.8 AIR

Earlier methods were based on polarography, spectophotometry of lead dithizonate and reaction with iodine monochloride (Moss and Browett[219]). None of these methods are of adequate sensitivity to be applicable to environmental air analysis. Methods based on gas chromatography combined with atomic absorption spectrometry operated with a lead specific detector are more appropriate to such measurements as discussed below.

**Atomic absorption spectrometry**

Various workers[220-222] have applied this technique to the determination of organolead compounds in air.

Methods such as direct atomic absorption spectrometry and flame photometry[222] are unsuitable for the determination of tetraalkyllead in environmental samples because of the lack of specificity and ion sensitivity or because of the long time required to analyse a single air sample.

Torsi and Palmisano[223] have described a procedure for sampling air using a battery powered field sampler and capturing organolead compounds by electrostatic attraction and subsequently determining them by electrothermal atomic absorption spectrometry. Down to 100 pg tetraalkyl lead (as lead) can be determined by this procedure.

**Gas chromatography**

Gas chromatography with electron capture detection and gas chromatography with catalytic hydrogenation prederivatisation and flame ionisation detection[224] and with microwave plasma detection[225] are unsuitable for tetraalkyl lead compounds in environmental samples because of the lack of specificity and/or sensitivity or because of the long time required for the analysis of a single air sample. Most of the available methods with high sensitivity and low detection limits are based on a sampling step, often performed by cryogenic trapping, followed by an analysis step by gas chromatography combined with atomic absorption spectrometric detection (both flame and electrothermal)[197,226,227].

These methods determine the tetraalkyllead compounds present but are time-consuming and require a complex set up.

Boettner and Dallas[228] have compared the sensitivities of the electron capture, thermal conductivity, argon ionisation and flame ionisation detectors in the chromatographic determination of organolead and aliphatic chloride and lead scavenging compounds in the atmosphere (such as are used in petroleum blends). They used a Wilkins Hi-Fi Model 600 Chromatograph with both hydrogen flame and electron capture detectors, a Beckmann Model GC-2A Chromatograph with a thermal conductivity detector and a Research Specialities Model 600 Chromatograph with an argon ionisation detector. All separations were made on 3 mm ($\frac{1}{8}$ in) or 6 mm × 2 mm ($\frac{1}{4}$ in × 6 ft) stainless steel columns. The compounds studied and the column coatings used are tabulated in Table 6.23.

The sensitivities of the thermal conductivity and the argon ionisation detectors are independent of the molecular weight or the number of chlorine atoms in the chlorinated compounds, but the flame detector decreases slightly in sensitivity

**Table 6.23.** Column coatings and supports for the separation of chlorinated aliphatic and lead alkyl compounds using gas chromatographs with various detector. (From Boettner and Dallas[228]. Reproduced by permission of United Trades Press, USA.)

|  | Thermal conductivity | Argon ionization | Flame ionization | Electron capture |
|---|---|---|---|---|
| Methyl chloride | C | A | A | A |
| Dichloromethane | C | A | C | B |
| Chloroform | C | A | A | B |
| Carbon tetrachloride | A | A | A | B |
| Ethyl chloride | A | A | A | A |
| 1,2 Dichloroethane | A | A | A | B |
| 1,1,1 Trichloroethane | C | A | A | B |
| 1,1,2 Trichloroethane | C | A | C | B |
| 1,1,2,2 Tetrachloroethane | C | A | C | E |
| 1,2 Dichloropropane | C | A | C | C |
| 1,2,3 Trichloropropane | C | A | C | E |
| Chloroethylene | A | A | A | A |
| 1,2 Dichloroethylene cis | A | A | A | A |
| 1,2 Dichloroethylene trans | A | A | C | B |
| Trichloroethylene | C | A | A | B |
| Tetrachloroethylene | D | A | C | B |
| Lead tetramethyl | D | G | D | F |
| Lead tetraethyl | D | G | F | F |

A — 20% Carbowax 600 on C-22 Firebrick
B — 5% Silicone 550 and 5% Ucon (Water insoluable) on Chromosorb P
C — 10% Silicone 550 on C-22 Firebrick
D — 20% Carbowax 20 M on C-22 Firebrick
E — 10% Silicone SE 30 on Chromosorb W
F — 5% Silicone SE 30 on Chromosorb W
G — 5% Silicone 550 on Anakrom ABS.

with increasing numbers of chlorine atoms. The electron capture detector was found to have its greatest response to the chlorinated compounds at 10 V. With the electron capture detector the sensitivity was dependent in rather a complex manner on the molecular weight and the number of chlorine atoms in the chlorine compound.

The conclusions reached by Boettner and Dallas[228] concerning the electron capture detector were as follows: that for analysis of volatile chlorinated aliphatic hydrocarbons this detector is no more sensitive than the ionisation detectors for those compounds having one or two chlorine atoms. For those compounds having three or four chlorine atoms, the electron capture detector was from 100 to 1000 times more sensitive than the ionisation detectors. For the two alkyl-lead compounds tested, the electron capture detector gives little improvement in sensitivity but its discrimination toward the lead substituted compounds as compared with unsubstituted hydrocarbons makes it a preferable detector for analysing mixtures of these two types of compounds.

Gas chromatography using an atomic absorption spectrometer as detector has been used by several workers to determine tetraalkyllead compounds[226,227,228,230].

Robinson et al.[230] have carried out a detailed study of atomisation processes in carbon furnace atomisers during the development of a gas chromatography-furnace atomic absorption combination for the determination of organolead compounds in gasoline and air. The carbon furnace atomiser is attached directly to the gas chromatograph column. The atomiser exhibits high sensitivity and eliminates many of the problems involved with interferences encountered with furnace atomisation. Their study of the furnace atomisation step revealed the problems involved in obtaining accurate quantitative data. Thus, it is vital to control all the variables which affect the rate and degree of atomisation of lead and to use pyrolytic carbon sample boats.

A further problem relates to the variation of the electrical resistance of the carbon rod or tube with use. Thus, with a standardised atomisation programme, the temperatures achieved will differ somewhat as the carbon ages, although the time and voltage remain the same, and this will affect the atomisation rate.

These difficulties lead Robinson et al.[230] to develop a new type of atomiser in which the effluent from the gas chromatograph enters the base of the atomiser where the gaseous sample is decomposed and the atomisation takes place. The atoms flow into the cross-piece which is in the optical light path. The advantage of the process is that the peak of the solvent used is quite separate from the peak of the metal-bearing component on the gas chromatogram. The gas chromatograph separates the metal-bearing components from the rest of the material, which eliminates many of the problems encountered in the solvent evaporation step and other matrix effects. Decomposition is fairly rapid, although several seconds elapse from the time that the sample enters the carbon atomiser before the atoms reach the optical light path. This permits chemical decomposition to take place and virtually eliminates chemical interference, which is usually caused by varying rates of atomisation from different compounds rather than by prevention of decomposition. Even if the rate varies, decomposition is virtually complete before the free atoms enter the light path. The peak height is a product of the metal concentration and the resolution of the g.c. column. The definition of sensitivity as that concentration which results in 1% absorption is therefore unrealistic. Sensitivity measurements were based on peak area data.

The sensitivity of the equipment was shown to be of the order 0.1 ng lead. Robinson et al.[230] studied the applicability of their technique to the determination of organolead compounds in the atmosphere. In this work 1 $m^3$ of air was pulled through a cryogenic trap. The trapped material was then put into the gas chromatograph under conditions suitable for tetramethyllead or other organic lead compounds. The concentrations of tetraethyllead were so small that in many samples none was detected.

De Jonghe et al.[226,227] have described a sampling system for the analysis by gas chromatography–atomic absorption spectrometry of alkyllead compounds in air. This method, when compared with many other published procedures, is relatively rapid.

Sampling periods of 1 h or less proved to be sufficient, even for the determination of alkyllead species in relatively nonpolluted air. The major difficulty in collecting the compounds from air samples on gas chromatographic column packing material is the condensation of moisture from the air sample in the trap. Ice condensation on the column material leads to clogging of the pores and a sharp decrease in the air flow rate. The volume of air that can be sampled is therefore limited.

De Jonghe et al.[226,227] used a large U-tube filled with glass beads at $-80°C$ to condense water from the sample before it entered the column.

In this way the predeposition of water was much improved as a result of a better cooling efficiency of the air. However, at $-80°C$ a substantial fraction of the tetraalkylleads is retained also, especially the less volatile species, whereas at higher temperatures also a rapid obstruction of the chromatographic adsorption tube occurs. Provided the trap is cooled down to sufficiently low temperatures to retain also the more volatile species, dimethyldiethyllead, trimethylethyllead and tetramethyllead it could therefore be used for the direct collection of the lead alkyl compounds. This would allow much higher air flow rates than is possible with chromatographic adsorption tubes.

The air to be analysed was passed at a flow rate of about 6 L min$^{-1}$ for 1 h through a two-component collection system. The first stage was a 47 mm Nucleopore membrane filter (0.4 µm) to remove the lead-containing particulates. The second stage was a cryogenic sampling trap for the collection of the volatile tetraalkyllead compounds. It consists of a U-shaped Pyrex tube (50 cm long by 25 mm i.d.) filled with glass beads of 4 mm diameter and immersed in a liquid nitrogen–ethanol slush bath at $-130°C$. After sampling is completed the U-tube remains in the slush bath until analysed.

After the sample was obtained the trapped alkyllead compounds were thermally desorbed from the large U-tube and transferred to a short adsorption tube by connecting the sampling tube, still immersed in the slush bath, with a short glass column (26 cm long by 6 mmo.d. and 2 mm i.d.) packed with 0.2 g of 3% OV-101 on 100/120 mesh Gaschrom Q and kept in liquid nitrogen. While air was passed at a flow rate of 1 L min$^{-1}$ the U-tube was removed from the slush bath and allowed to warm slowly in air and then in a water bath, at 60°C. With this treatment a rather constant air flow through the desorption system can be maintained, but near the end of the operation, when appreciable amounts of water start to evaporate out of the trap and condense on top of the adsorption tube, the flow rate decreases and desorption stops.

The adsorption tube was then attached to the four-port valve installed between the carrier gas inlet and the injection port of the gas chromatograph. The tube was immersed in a hot water bath at ca. 90°C, and the trapped sample was swept into the gas chromatograph by the carrier gas. Simultaneous with this injection, the gas chromatograph oven temperature program was initiated and the graphite furnace brought to 2000°C.

The reproducibility of measurement achieved by this method are better than 2% at the nanogram level. A detection limit of below 100 pg lead m$^{-3}$ air was achieved. Typical values of the five ethyl methyl lead compounds in suburban and residential air samples ranged between <0.02 and 30.3 µg m$^{-3}$.

Radzuick et al.[229] have described a sensitive gas chromatographic graphite furnace atomic absorption method for the determination of individual alkyllead compounds in air. The atomic absorption instrument was operated at the 283.3 nm leadline. Gas chromatography–mass spectrometry was used to identify the separated organolead compounds.

A Perkin-Elmer 603 atomic absorption spectrophotometer was equipped with a deuterium background corrector, and a HGA 2100 graphite furnace. The radiation source was a Perkin-Elmer electrodeless discharge lamp operated at 10W. A Pye series 104 chromatograph was interfaced with the graphite furnace with a tantalum connector machined from a 6.4 mm diameter rod. The glass chromatographic column (150 cm long, 0.6 cm o.d.) was packed with 3% OV-101 on Chromosorb W operated at 150°C. The effluent was transferred to the furnace by Teflon-lined aluminium tubing (3 mm o.d.) heated electrically to 80°C. The adsorption tube for air samples were U-shaped Teflon-lined aluminium tubes (30 cm long, 3 mm o.d.) packed with 3% OV-101 on Chromosorb W (80–100 mesh). Moisture was condensed from air by using glass U-tubes at −15°C.

The detection limit of this method was found to be about 40 pg of lead for each compound, based on peak-height measurements. For a 70 L air sample, 0.5 ng m$^{-3}$ of each compound could be detected.

Radzuick et al.[229] showed that total atmospheric alkyllead averaged 14 ng Pb m$^{-3}$ in the samples they examined.

Vehicular exhaust fumes are an insignificant contributor to this total. Tetraethyllead, the only alkyllead compound used in Southern Ontario gasoline, is unstable in air. Besides decomposing, it reacts to give other alkyllead compounds, which can also be determined.

Reamer et al.[225] have discussed the applicability of a gas chromatograph coupled with a microwave plasma detector for the determination of tetraalkyllead species in the atmosphere. The tetraalkyllead species are collected by a cold trap. The volatile lead species are concentrated within an organic solvent, separated by a gas chromatographic column and determined by measurement of emission intensity at the lead 405.78 nm line. A cold trap containing SE-52 on Chromosorb F at −80°C was used to collect alkyllead compounds from the atmosphere. This trap had a collection efficiency for tetraalkyllead compounds of 84–100% at the 150 µg tetraethyllead level.

When using the wavelength modulation mode for background correction, the tetraalkylead calibration curves extended from the low pg range to the low ng range with the following detection limits: tetramethyllead, 6 pg; trimethylethyllead, 10 pg; dimethyldiethyllead, 23 pg; methyltriethyllead, 35 pg; and tetraethyllead, 40 pg.

The absorbent from the sampling tube was transferred to a 50 mL round-bottom flask which was attached to a miniature freeze-drying system. The sample was dried for 12 h, resulting in the quantitative removal of the tetraalkyllead compounds and water present in the sample. The gases were quantitatively trapped cryogenically at liquid nitrogen temperature. The trap was removed from the system and slowly warmed. The container walls were rinsed with distilled water to ensure quantitative retention of the analyte. The organic lead compounds were extracted with 200 µL of hexane by shaking for 10 min. Sample analysis involved injecting a 1 to 5 µL aliquot of the hexane layer into the gas chromatograph with a 10 µL microsyringe.

Individual concentrations of the five ethyl/methyl lead compounds found in vehicle exhausts ranged between 2 and 650 µg m$^{-3}$ and total tetraalkyllead compounds ranged from 57 to 1030 µg m$^{-3}$.

## High performance liquid chromatography

Koizumi et al.[231] coupled a Zeeman graphite furnace atomic absorption detector to a high performance liquid chromatographic column to separate alkyllead compounds collected from air samples and automotive exhausts. One of the problems with conventional graphite furnace atomic absorption spectrometers is that, because the maximum attainable atomisation temperature is too low different absorbance values are obtained for each of the five ethyl methyl lead compounds. One way of overcoming this difficulty is to atomise at a very high temperature. Koizumi et al.[231] achieved this with a new detector design which is capable of achieving an atomisation temperature of 2800°C. This furnace consists of several separate parts; the sample cup, the thermal converter and reactor of porous graphite, a narrow hole (which at high temperatures also acts as a thermal converter), and the absorption cell. The sample vapour flows through the thermal converter and its temperature is raised sufficiently to decompose the compound and to atomise the metal. After that, the sample vapour is carried to the absorption cell for Zeeman atomic absorption spectrometry measurement.

The centre portion of the cuvette is heated to a high temperature first because of the small heat capacity and the large electrical resistance. Conductive heating causes the porous graphite to be heated next and, finally, the tantalum cup is heated. A few seconds after the current is turned on, the temperature difference between the three sections begins to decrease. Vaporised sample in the cup follows the flowing argon gas through the porous graphite (the temperature of which is raised by coming into intimate contact. It then passes through a small hole, the walls of which have the highest temperature). Because of the intimate contact with these surfaces, the gas temperature becomes equal to the wall temperature before passing into the absorption cell.

The system used to separate alkyllead compounds is described in Table 6.24 Methyl alcohol was used as the eluent. The pressure was about 30 kg cm$^{-2}$, and the flow rate was 0.67 mL min$^{-1}$. A sample of 10 mL was injected into the high performance liquid chromatograph while the flow was stopped. A 10 µL aliquot from each 250 µL portion of column effluent was intermittently introduced into the furnace.

Table 6.24. Instrumentation for HPLC-ZAA system. (From Koizumi et al.[231] Reproduced by permission of American Chemical Society.)

HPLC
   high performance liquid chromatograph Hitachi M633,
      0.350 kg cm$^{-2}$, 0.36 3.6 mL min$^{-1}$.
   column. Hitachi, 2.5 × 500 mm
   resin, Hitachi Gel No. 3010
Furnace
   graphite. Ultra Carbon 0.5 - in. diameter,
   porous graphite, RVC 100 PPI porosity grade
   furnace power supply, reactor controlled, 20 V. 700 A
ZAA spectrophotometer.
   light source, magnetically confined lamp (dc + rf (50 Mhz)).
   magnet, permanent 12 kg
   variable retardation plate, 0/2, 30 Hz
   polariser, Rochon prism (quartz optical contact)
   monochromator, Hitachi M100 spectrophotometer
   photomultiplier, Hamamatsu T.V. YA 7122
   chopper, Bulova L2C, 1.0 KHz
electronics.
   lock-in amplifier, (including log convertor, AGC)
   recorder, Honeywell Electronik 17.

To determine organolead compounds in automotive exhaust gases, the gas was collected in a polythene bag and was forced to flow through the furnace carrier gas inlet port at a flow rate of 0.09 L min$^{-1}$. The lead concentration in the exhaust gas was determined from the area under the absorption signal.

## 6.3 ORGANOTIN COMPOUNDS

### 6.3.1 NATURAL AND POTABLE WATERS

**Spectrophotometric methods**

Luskima and Syatsillo[232] have described a spectrophotometric procedure utilising phenylfluorone for the determination of organotin compounds in water. They also used gas chromatography to separate tetraethyltin and tetrabutyltin.

**Spectrofluorimetric methods**

Spectrofluorometry has been applied to the determination of triphenyltin compounds. Coyle and White[233] showed that 3-hydroxyflavone could be used to determine submicrogram amounts of inorganic tin and then Vernon[234] used the reagent to determine triphenyltin compounds in potatoes. On the basis of this procedure Blunden and Chapman[235] spectrofluorometrically determined triphenyltin compounds in water. Further, they showed that chloride ions quenched the fluorescence but that on shaking with aqueous sodium acetate solution a stable complex was formed, although the instability to light of the triphenyltin chloride-3-hydroxyflavone complex had been initially pointed out by Aldridge and Cremer[236]. Triphenyltin compounds in water at concentrations of 0.004–2 mg L$^{-1}$ are readily extracted into toluene and can be determined by spectrofluorimetric measurements of the triphenyltin-3-hydroxyflavone complex. Tri-, di-, and monobutyl and di- and monomethyltin compounds did not fluoresce under the conditions used for the determination of triphenyltin. However, trimethyltin compounds react in a similar manner with 3-hydroxyflavone, and although the emission maximum is at approximately 510 nm, this is not sufficiently different from the emission maximum of triphenyltin compounds (approximately 495 nm) for these compounds to be determined in the presence of each other.

Spiking recoveries by the above procedure carried out on standard solutions of triphenyltin chloride in various types of water ranged from 74% at the 4 μg L$^{-1}$ tin level (rel. s.d. 8.9%) to 93.6% at the 2 mg L$^{-1}$ level (rel. s.d. 4.2%).

## Gas chromatography with various detectors

Chau et al.[237] described an extraction procedure for the polar methyltin compounds and the use of the gas chromatography atomic absorption spectroscopy system for the determination of their butylated derivatives including inorganic tetravalent tin. Butylation of the organotin compounds was carried out with butyl magnesium chloride reagent. This procedure meets even the most searching present day requirements as regards to speciation and sensitivity in the determination of organotin compounds.

Large volumes of water sample can be handled. Under normal laboratory conditions, detection limit of 0.04 μg L$^{-1}$ can be achieved with 5 L of water sample. The absolute detection limit of the GC–AAS for tin is 0.1 ng. Volatile organotin compounds such as tetramethyltin and methyltin hydrides can also be analysed by this method.

In this method, the highly polar and solvated methyltin, dimethyltin, trimethyltin, and Sn(IV) species were extracted into benzene containing tropolone from water saturated with sodium chloride.

After the mixture was shaken for 0.5 h the benzene layers were separated and butylated with 1 mL of butyl magnesium chloride reagent in 10 mL glass stoppered micro Erlenmeyer flasks with stirring for ca. 10 min. The mixture was washed with 5 mL of 1 N sulfuric acid or hydrochloric acid to destroy the excess Grignard reagent. The organic phase was separated and dried with anhydrous sodium sulfate. The mixture now contained the butyl-derivatised methyltins, $MeSnBu_3$, $Me_2SnBu_2$, $Me_3SnBu$ and $Bu_4Sn$ ready for analysis by the gas chromatography-atomic absorption technique. The addition of solid sodium chloride to the original water sample improved recoveries of organotin compounds.

The absolute recovery was found satisfactory for $Me_2Sn^{2+}$, and $MeSn^{3+}$ and Sn(IV), but was only 75% for the $Me_3Sn^+$ species even with 40 g of sodium chloride used to achieve a saturated salt solution. The $Me_2Sn^+$ species was only extracted in the presence of sodium chloride. The recovery, although not quantitative, was consistent.

Chau et al.[237] found that the introduction of hydrogen to the quartz furnace was necessary to elevate the furnace temperature to ca. 900°C and to enhance atomisation of the methylbutylin derivatives. It was also found that introduction of air further enhanced the sensitivity. Heating the transfer line to 165°C was necessary to give sharp peaks. Although this temperature was well above the boiling points of the tin derivatives no decomposition of the alkyltin compounds was noted at the transfer line. Precision at the 250 µg $L^{-1}$ level of $Me_3Sn^+$, $Me_2Sn^{2+}$, $MeSn^{3+}$ and Sn(IV) in lake water was acceptable as reflected in standard deviations of 5.4, 8.6, 7.3 and 11%, respectively and recoveries of all species were in the range 91–117%.

The only tetravalent elements that were coextracted by tropolone and similarly butylated to the tetraalkyl derivatives are the germanium(IV) and lead(IV) species. Tetraalkyllead compounds, however, do not give any signals in the atomic absorption detection system for tin analysis at the 224.6 nm spectral line. There should not be any worry of interference from germanium(IV) in natural waters nor is its spectral interference expected.

Tetramethyltin and organotin hydrides can be analysed by direct gas chromatography–atomic absorption spectrophotometry without derivatisation. As these compounds are volatile it is necessary to purge them from the water sample and the headspace of the sample vessel with nitrogen into U-trap packed with 3% OV-1 on Chromosorb W at 160°C and subsequently mounting this trap onto the inlet of the gas chromatographic column.

Between 0.1 and 0.5 µg $L^{-1}$, $Me_2Sn^{2+}$, $MeSn^{3+}$ and Sn(IV) were found in harbours, or industrialised areas by this method ($Me_3Sn^+$ not detected).

Hodge et al.[238] have described an atomic absorption spectoscopic method for the determination of butyltin chlorides and inorganic tin in natural waters, in amounts down to 0.4 ng.

Various solvents have been used to preconcentrate organotin compounds from natural waters including benzene–tropolone[237,239], n-hexane-tropolone[240], methylene dichloride[241] and benzene[242].

Neubert and Wirth[243] report on the quantitative determination of mono-, di-, tri-, and tetraalkyltin compounds, present in a mixture by gas chromatography after alkylation of the mixed tetraalkyltins. This technique was applied by Neubert and Andreaa[244] to the quantitative detection of tributyl- and dibutyltin species present in dilute aqueous solution. Butyltin species were concentrated on a cation-exchange column, desorbed into diethylether–hydrogen chloride and determined by gas chromatography after methylation.

Gas chromatographic–atomic absorption spectrometric methods for the determination of nanogram amounts of methyltin compounds and inorganic tin in natural waters and human urine have been described by Braman and Tomkins[245]. In this method the tin compounds in aqueous solution at pH 6.5 are converted by sodium borohydride to the corresponding volatile hydride, $SnH_4$, $CH_3SnH_3$, $(CH_2)$-$SnH_2$ and $(CH_3)_3SnH$ by reaction with sodium borohydride. These are helium scrubbed from solution, cryogenically trapped on a U-tube and separated upon warming. Detection limits are approximately 0.01 ng as tin when using a hydrogen-rich, hydrogen-air flame emission type detector ($S_nH$ band) of a type having very low detection limits. Average tin recoveries ranged from 83–108% for six samples analysed to which were added 0.4 to 1.6 ng of methyltin compounds and 3 ng inorganic tin. Reanalysis of analysed sampled shows that all methyltin and inorganic is removed in one analysis procedure.

Braman and Tomkins[245] used a gas chromatographic column consisting of fully packed 30 cm U-tubes of silicone oil type OV-3, on Chromosorb W to separate the stannanes. They noted a resolution of two peaks within the dimethylstannane signal. Similar results were observed during the analysis of natural waters containing dimethyltin compounds. This may be due to the formation of stable bipyramidal geometric isomers of dimethylstannane but this was not verified. While the reasons for the observed effect are not known, the split signal is quantitative for dimethyltin dichloride and produced no problems during analysis. Detection limits are in the range 0.07–0.2 ng $L^{-1}$ depending on the compound detected. The precision of the method averaged ± 5% relative over the range of the response curve.

Braman and Tomkins[245] did not observe any interference by organics in this method in the determination of organotin compounds in water samples. They did, however, observe that organic arsenic(III) compounds in water caused a positive interference effect due to the emission of the As $^{2+}$ molecular band at 611.5 nm. Inorganic arsenic(III) is reduced to arsine at pH 6.5. It is, nevertheless, separated from stannane on the OV-3 column and is not an interference. The arsenic(III) peak can be eliminated by oxidation of arsenic(III) to arsenic(V) by the addition of a few drops of sodium thiosulphate solution to dispel excess iodine. Neither arsenic(V) nor the methylarsenic acids are reduced to corresponding arsines at pH 6.5.

Certain metal ions, $Ag^+$, $Cu^{2+}$, $Hg^{2+}$, $Ni^{2+}$, $MoO_4^{2-}$ and $Pb^{2+}$ at 2 µg $L^{-1}$ in analysed solutions were found to reduce the complete removal of stannane. Sea water did not inhibit recovery of stannane or methylstannanes. The ions $Al^{3+}$, $CrO_4^{2-}$, $F^-$, $I^-$, $Mn^{2+}$, $PO_4^{3-}$ and $Sb^{3+}$ did not interfere at 20 µg $L^{-1}$ while $Fe^{3+}$, $BiO_3^-$, $Cd^{2+}$, $S^{2-}$, $VO_3^{2-}$ and $Zn^{2+}$ did not interfere at 2 µg $L^{-1}$.

Braman and Tomkins[245] present extensive analytical data for the four methyltin species in saline, and estuarine waters, surface waters and rain waters obtained at a variety of locations in the United States (see Section 6.3.2).

Chau et al.[237] have described an improved extraction procedure for polar methyltin compounds, using benzene containing tropolone, from water saturated by sodium chloride. Tetramethylbutyltin derivatives were prepared in the extracts and were separated by gas chromatography in well-defined peaks. The difference in sensitivity of the different tin species is attributed to differences of behaviour in the atomic absorption furnace. The overall recovery is satisfactory, coefficients of variation using six replicate samples was 5–11% and a detection limit of 0.04 µg L$^{-1}$ was achieved.

Soderquist and Crosby[241] have developed a method for the simultaneous determination of triphenyltin hydroxide and its possible degradation products tetraphenyltin, diphenyltin oxide, benzenestannoic acid (and inorganic tin) in water. The method is rapid (one sample set per hour), sensitive to less than 10 µg L$^{-1}$ for most of the tin species and exhibits no cross-interferences between the phenyltins. The phenyltins are detected by electron capture gas-liquid chromatography after conversion to their hydride derivatives, using lithium aluminium hydride, while inorganic tin is determined by a spectrophotometric procedure which responds to tin(IV) oxide as well as aqueous tin(IV).

Soderquist and Crosby[241] found that nonvolatile hydroxyoxphenyl stannane (PhSnO$_2$H), oxodiphenyl stannane (PhSnO) and hydroxy-triphenyl stannane (Ph$_3$SnOH) upon conversion to their hydrides by lithium aluminium hydride produced derivatives with excellent gas chromatographic properties, high response to electron-capture detection and none of the attendant column stability problems encountered with other derivatives.

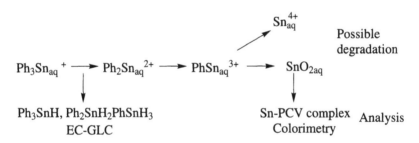

The basis for this method involves extraction of the phenyltin species from water with dichloromethane followed by their quantification as phenyltin hydrides by electron capture gas chromatography and analysis of the remaining aqueous phase for inorganic tin (Sn$^{4+}$ plus SnO$_2$) by colorimetry.

Soderquist and Crosby[241] used a dual column/dual detector Varian model 2400 gas chromatograph equipped on one side with a flame-ionization detector and a 0.7 m by 2 mm (i.d.) glass column containing 3% OV-17 on 60/80 mesh gas Chrom Q. Column, injector and detector temperatures were 265, 275 and 300°C, respectively, carrier gas (nitrogen) flow rate was 25 mL min$^{-1}$. Tetraphenyltin eluted within 8 min under those conditions. The second side of the chromatograph was equipped with a tritium EC detector and a 1.1 m by 2 mm (i.d.) glass column containing 4% SE-30 on 60/80 mesh Gas Chrom Q. The injector and detector temperatures were 210°C and the carrier gas (nitrogen) flow rate was 20 mL min$^{-1}$. Column temperatures which eluted the following compounds within 6 min were: triphenylstannane (PnSnH) (190°C), diphenylstannane ($Ph_2SnH_2$) (135°C) and phenylstannane ($PhSnH_3$) (45°C).

None of the natural water samples analysed by Soderquist and Crosby[241] contained materials which interfered with the determination of any of the tin compounds of interest.

## Organotin compounds and elemental tin

## Method

### Organotin compounds

### Reagents

Unless otherwise noted, water was distilled and passed through a column of Amberlite XAD-4 resin. Hexane and dichloromethane were Nanograde quality or equivalent. All glassware was cleaned by soaking in 2.0 M hydrochloric acid followed by rinsing with copious volumes of water.

*Lithium aluminium hydride solution*, add 100 ± 10 mg solid lithium tetrahydridoaluminate to 25 mL of dry, reagent grade diethylether in a glass-stoppered flask. Shake for 2 min and allow the grey precipitate to settle before use. Prepare the solution fresh daily.

*Sensitised PCV solution*, Pyrocatechol violet (Eastman Organic Chemicals), add 12 mg solid and 11 mg of cetyltrimethyl ammonium bromide to 1 L of water. Prepare fresh daily.

*Ascorbic acid solution*, Add 225 g solid to 50 mL of water. Prepare fresh daily.

*Acetate buffer* (2.0 M, pH 4.7). Combine sodium acetate trihydrate (68 g) and 28 mL of glacial acetic acid and dilute to 500 mL with water.

*Citric acid solution*, add 10 g solid citric acid monohydrate to 100 mL of water.

*Sulphuric/citric acid solution*, Combine analytical reagent grade sulphuric acid (25 g) and 13 g of citric acid monohydrate and dilute to 500 mL with water.

### Procedure (Figure 6.8)

### Organotins

Mix 200 mL sample in a 250 mL separatory funnel with 5 mL of acetate buffer, extract the mixture with two 15 mL portions of dichloromethane and divide the pooled extract into three equal parts. Concentrate each to about 0.1 mL in a screw-capped test tube at less than 40°C under a gentle stream of nitrogen.

To one of the concentrates (EX-1, Figure 6.8), add 5 mL of hexane followed by 0.5 mL of lithium aluminium hydride solution. After 2–3 min, dilute the mixture with hexane add about 0.5 mL of water, mix the phases and analyse the hexane phase by electron capture gas chromatography for $Ph_3SnH$, $Ph_2SnH_2$, and $PhSnH_3$, Prepare a standard curve for each of the hydrides using ng $\mu L^{-1}$ hexane standard solutions, generally in the 0.2–2.0 ng range. Achieve quantitation by comparison of sample peak hights to the standard curve.

**Tetraphenyltin**

To the second dichloromethane concentrate (EX-3 in add Figure 6.8) add 1 mL of hexane, concentrate the contents under nitrogen to about 0.1 mL and then dilute back to 1.0 mL with hexane. Transfer the sample to a Florisil microcolumn (prepare by packing a disposable Pasteur pipette with 0.35 g of 60–100 mesh Florisil held with a small glass wool plug and rinse with two 5 mL portions of hexane before use), and elute with hexane. Collect the first 2.5 mL of eluate, concentrate to 0.1–0.5 mL and analyse by flame ionisation gas chromatography for tetraphenyltin. Achieve quantitation by comparison of sample peak heights to the tetraphenyltin standard curve in the 10–50 ng range.

**Inorganic tin**

To the second dichloromethane concentrate (EX-2) referred to in Figure 6.8 add 0.50 mL of sulphuric acid and remove the dichloromethane from the mixture with a vigorous stream of nitrogen. Seal the tube and heat at 100°C in a water bath for 20 min. After cooling, add 4.0 mL of citric acid solution and treat the sample as described below.

Divide the extracted aqueous sample equally between two 125 mL Erlenmeyer flasks and concentrate the samples by boiling just to dryness on a hot plate. To one of the dry aqueous concentrates (AQ-1, Figure 6.8) add 2.0 g of potassium

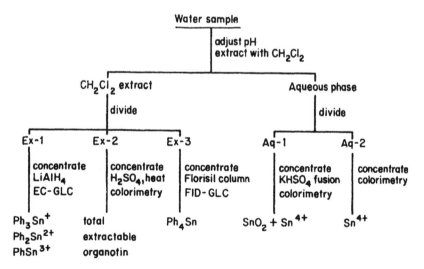

**Figure 6.8.** Flow diagram for analytical procedure. (From Soderquist and Crosby[241]. Reproduced by permission of American Chemical Society.)

hydrogen sulphate and heat the flask at 300–350°C for 30 min. After cooling, add 4.0 mL of citric acid solution and dissolve the solids with gentle heating if necessary. Remove any insoluble particulate matter which would interfere with subsequent colorimetric measurements by filtering through tightly packed glass wool. To the second aqueous concentrate (AQ-2), add 4.0 mL of sulphuric acid–citric acid solution. Treat both samples as described below under colour development.

**Colour development**

Analyse the hydrolysed dichloromethane extract (EX-2) and the two aqueous samples (AQ-1 and AQ-2) for tin as follows: add ascorbic acid solution (2.0 mL) and 4.0 mL of sensitised pyrocatechol violet solution and after 30 min read the absorbance at 660 nm.

Prepare a standard curve for the aqueous, tin subsamples (AQ-1 and AQ-2) by the addition of 0, 1.0, 3.0, and 5.0 μg of tin to 3.5 mL of sulphuric citric acid solution followed by 2.0 mL of ascorbic acid solution, 4.0 mL of sensitised pyrocatechol violet solution and enought water to give 11.5 mL total volume. Prepare a separate standard curve for the hydrolysed dichloromethane extracts, (EX2) when the final sample volume, after colour development, was less than about 15 mL; in these cases addition of the same tin, standards was made to 0.5 mL of sulphuric acid plus 4.0 mL of citric acid, again followed by the above amounts of ascorbic acid and sensitised pyrocatchecol violet.

Hattori et al.[242] determined trialkyltin and triphenyltin compounds in environmental water samples. The water samples were mixed with hydrochloric acid and sodium chloride and extracted into benzene. Following dehydration and concentration, the compounds were cleaned up, using glass columns packed with silica gel impregnated with hydrochloric acid and their hydrides generated using an ethanol solution of sodium borohydride. The organotin hydrides were determined using gas chromatography with electron capture detection. Recoveries were 7–95% from river water and the detection limit 0.8 μg L$^{-1}$.

Jackson et al.[250] devised trace speciation methods capable of ensuring detection of tin species along with appropriate preconcentration and derivatisation without loss, decomposition, or alteration of their basic molecular features. They describe the development of a system employing a Tenax GC filled purge and trap sampler, which collects and concentrates volatile organotins from water samples (and species volatilised by hydrodization with sodium borohydride) coupled automatically to a gas chromatograph equipped with a commercial tin selective flame photometric detector modified for tin-specific detection[247–249].

Lobinski et al.[250] optimised a method for the comprehensive speciation of organotin compounds in environmental samples such as water and sediments. Sample components were separated by capillary gas chromatography and detected by helium microwave induced plasma emission spectrometry. Mono-, di-, tri- and some tetra-alkyltin compounds were resolved. The ionic organotin compounds were extracted as their diethylthiocarbamates into pentane then converted to pentyl magnesium bromides for gas chromatography. A detection limit of 0.05 pg tin was achieved. These workers found 33, 18 and 14 ng L$^{-1}$, respectively, of tri-, di- and monobutyltin in environmental waters.

## Gas chromatography–mass spectrometry

Meinema et al.[239] have described a sensitive and interference free method for the simultaneous determination of tri-, di- and monobutyltin species in aqueous systems at tin concentrations of 0.011–5 µg L$^{-1}$. The species are concentrated from hydrobromic acid solutions into an organic solvent by extraction with tropolone in the presence of a metal coordinating ligand. The butyltin species in the organic extract are transformed into butylmethyltin compounds by reaction with a Grignard reagent and analysed by a gas chromatography mass spectrometry method. The inorganic tin(IV) species in the organic extract are butylated to tetrabutyltin which is detected by the same technique.

One litre of this aqueous solution, acidified with 20 mL of hydrobromic acid (48%), was extracted twice with 25 mL of benzene containing 0.05% tropolone. The methylated benzene extract concentrated to a volume of 10 mL was submitted to the gas chromatography–mass spectrometry detection procedure. Butylmethyltin compounds were detected in 95% (Bu$_3$SnMe), 90% (Bu$_2$SnMe$_2$) and 84% (BuSnMe$_3$) yields calculated on the amounts of (Bu$_3$SnCl), Bu$_2$SnCl$_2$ and BuSnCl$_3$ originally present.

Tetramethyltin formed by the methylation of inorganic tin(IV) cannot be determined by gas chromatography–mass spectrometry since it has the same retention time as the solvent. Butylation of an organic extract (50 mL) that contains inorganic Sn(IV) species by the addition of an excess of butyl magnesium bromide in diethylether (4 mL of a 2.0 N solution) results in the formation of tetrabutyltin. The reaction mixture is stirred for 0.5 h at room temperature and subsequently treated with 25 mL of a 1 N sulphuric acid solution. The aqueous layer is separated and extracted with 25 mL of benzene, and the combined organic layers are concentrated at reduced pressure to a volume of about 25 mL. The tetrabutyltin content of this sample can be determined quantitatively by gas chromatography–mass spectrometry.

A comparison of the mass spectra of Bu$_n$SnMe$_{(4-n)}$ ($n = 1-3$) compounds reveals that they all show a common fragment ion $m/e = 135$. Moreover, the mass spectra of Bu$_2$SnMe$_2$ and Bu$_3$SnMe show a common fragment ion $m/e = 193$ (Figure 6.9). This implies that gas chromatographic mass spectrometric analysis of these compounds can be performed by multiple peak scanning of these two fragments (mass fragmentography). This procedure has advantages over the procedure of detecting the total ion current in that it is considerably more sensitive, whereas no or substantially less disturbance by impurities with the same retention time as the Bu$_n$SnMe$_{(4-n)}$ species occurs. A contaminant with the same retention time as one of the Bu$_n$SnMe$_{(4-n)}$ compounds is not detected at all unless it presents the same mass fragments, $m/e = 135$ and also $m/e = 193$. The presence of such a contaminant is easily observed because the mutual ratio of these fragments will be disturbed. When this is the case, it is possible to monitor fragment ions containing one of the other isotopes of tin. However, this will affect the sensitivity as can be seen from the spectra presented in Figure 6.9.

Quantitative determination is made possible by addition to the sample of an internal standard. Such an internal standard should have a mass fragments at $m/e = 135$ and $m/e = 193$. Moreover, it should display a retention time in the gas chromatograph different from those of the $Bu_n SnMe_{(4-n)}$ ($n = 1-3$) compounds. As shown in Figure 6.9 $HexBu_2SnMe$ meets these requirements.

Combined gas-liquid chromatography–mass spectrometry was performed on a Finnigan Model 1015 instrument utilising a 1.0 m by 2 mmm (i.d.) glass column containing 3% OV-17 on 60–80 mesh Gas Chrom Q. Infrared spectra were obtained in hexane solution.

Whilst recoveries of tri-and diphenyltin compounds were good, those of monophenyltin compounds were in the range 11 to 81%. Minimum detectable amounts (200 mL sample) ranged from 15 µg L$^{-1}$ ($Ph_4Sn$) to 3 µg L$^{-1}$ ($Ph_2Sn^{2+}$ and $PhSn^{3+}$ Table 6.25)

Typical chromatograms of the hydride standards near the limit of detectability are shown in Figure 6.10.

Matthias et al.[251] described a comprehensive method for the determination of aquatic, butyltin and butylmethyltin species at ultratrace levels, using simultaneous sodium borohydride hydridization/dichloromethane extraction with gas chromatographic flame detection and gas chromatographic–mass spectrometric detection. The detection limits for a 100 mL sample were 7 ng of tin L$^{-1}$ for tetrabutyltin and tributyltin, 3 ng of tin L$^{-1}$ for dibutyltin and 22 ng tin L$^{-1}$ for monobutyltin. For 800 mL samples detection limits were 1–2 ng tin L$^{-1}$ for tri- and tetrabutyltin and below 1 ng tin L$^{-1}$ for dibutyltin. The technique was applied to the detection of biodegradation products of tributyltin in natural waters. It was a rapid and simple analysis suitable for large scale environmental monitoring programs.

Maguire and Huneault[252] developed a gas chromatographic method with flame photometric detection for the determination of bis(tri-$n$-butyltin) oxide and some of its possible dialkylation products in potable and natural waters. This method involves extraction of bis (tri-$n$-butyltin) oxide, $Bu_2Sn^{2+}$, $BuSn^{3+}$ and $Sn^{4+}$ from the water sample with 1% tropolone in benzene, derivatisation with a pentyl Grigand reagent to form the various $Bu_n$ pentyl (4-$n$) Sn species followed by analysis of these by flame photometric gas chromatography–mass spectrometry in amounts down to 25 ng. The pentyl derivatives are all sufficiently non-volatile compared with benzene that none are lost in solvent 'stripping', yet they are volatile enough to be analysed by gas chromatography.

Mueller[253], detected tributyl tin compounds at trace levels in water (and sediments, see Section 6.3.4) using gas chromatography with flame

ANALYSIS OF ORGANOMETALLIC COMPOUNDS IN THE ENVIRONMENT 175

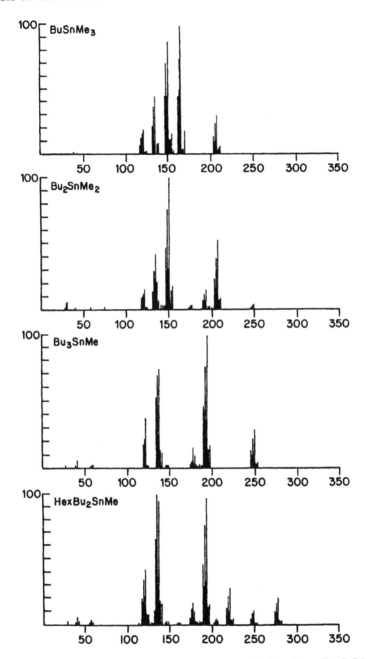

**Figure 6.9.** Mass spectra of $BuSnMe_3$, $Bu_2SnMe_2$, $Bu_3SnMe$ and hex $Bu_2SnMe$. (From Meinema et al.[239]. Reproduced by permission of American Chemical Society.)

**Table 6.25.** Method sensitivity. (From, Meinema et al.[239] Reproduced by permission of American Chemical Society.)

| Species | Method | Minimum detectable amount | Method sensitivity ($\mu g\ L^{-1}$)* |
|---|---|---|---|
| $Ph_4Sn$ | FID-g.l.c. | 5.0 ng as $Ph_4Sn$ | 15 |
| $Ph_3Sn^{1+}$ | e.c.-g.l.c. | 0.2 ng as $Ph_3SnH$ | 3 |
| $Ph_3Sn^{2+}$ | e.c.-g.l.c. | 0.2 ng as $Ph_2SnH_2$ | 3 |
| $PhSn^{3+}$ | e.c.-g.l.c. | 0.2 ng as $PhSnH_3$ | 3 |
| Total extractable organotins | colorimetry | 1.0 µg as Sn | 10 |
| $Sn^{4+}$ | colorimetry | 1.0 µg as Sn | 7 |
| $SnO_3 + Sn^{4+}$ | colorimetry | 1.0 µg as Sn | 7 |

*For 200 mL samples.

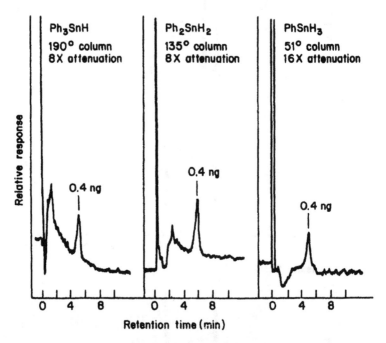

**Figure 6.10.** Typical chromatograms of the hydride derivatives. (From Meinmea et al.[239]. Reproduced by permission of American Chemical Society.)

photometric detection and gas chromatography–mass spectrometry. The tributyltin compounds are first converted to tributylmethyltin and then analysed using capillary gas chromatography with flame photometric detection and gas chromatography–mass spectrometry. Tributyltin was found in samples of river and lake water, and sediment and these results demonstrated the technique has detection limits of less than 1 pg $L^{-1}$.

Unger et al.[240] determined butyltin compounds in estuarine waters as their hexyl derivatives using gas chromatography with flame photometric detection. Traces of organotin compounds were analysed by solvent extraction with tropalone in $n$-hexane and derivatisation with $n$-hexyl magnesium bromide to form tetra-alkyltins which were then determined by gas chromatography with flame photometric detection and confirmation by mass spectrometry. The $n$-hexyl derivatives of methyltin and butyltin species were easily separated and quantified relative to an internal standard (tripentyltin chloride) which was not found in environmental samples and did not interfere in the method.

Muller[254] used high resolution gas chromatography with flame photometric detection to determine trace levels of organotin compounds in environmental water samples. Butyltin and other organotin compounds were determined via extraction, ethylation and capillary column gas chromatography. Identities were confirmed by mass spectrometry.

Colby et al.[255] used laser ionisation gas chromatography–mass spectrometry to determine tetraethyltin in natural water with a detection limit of 2.5 µg L$^{-1}$ as Et$_4$Sn or 1.5 fg absolute as Et$_4$Sn.

### High performance liquid chromatography

Nygren et al.[256] interfaced on-line a liquid chromatograph to a continuously heated graphite furnace atomic absorption spectrometer to determine di- and tributyltin species in natural waters with a detection limit of 0.5 µg tin absolute.

### Supercritical fluid chromatography

Shen et al.[257] evaluated indirectly coupled plasma mass spectrometry as an element detector for the supercritical fluid chromatography of organotin compounds in water. Detection limits of 0.04 and 0.047 pg absolute were obtained, respectively, for tetrabutyltin and tetraphenyltin.

High performance liquid chromatography coupled with hydride generation–direct current plasma emission spectrometry has been used for trace analysis and speciation studies of methylated organotin compounds in water[258].

Total tin was determined by continuous online hydride generation followed by direct current plasma emission spectroscopy. Interfacing the hydride generation–DC plasma emission spectrometric system with high performance liquid chromatography allowed the determination of tin species. Detection limits, sensitivities, and calibration plots were determined.

### Thin layer chromatography

The presence of Bu$_3$Sn and Bu$_2$Sn species in chloroform or benzene extracts from hydrobromic acid acidified natural waters can be demonstrated qualitatively by thin-layer chromatography on Eastman chromatogram sheets using hexane–acetone-acetic acid 40:4:1 as an eluant[259]. After spraying with a 0.1% dithizone solution in chloroform, $Bu_3Sn$ species are visualised as a yellow spot ($R_f$ 0.75) and $Bu_2Sn$ species as a red spot ($R_f$ 0.50). The detection limit is improved by keeping the thin-layer strip for 10 s in bromine vapour after elution. Bromine breaks down carbon–tin bonds, and both $Bu_3Sn$ and $Bu_2Sn$ species are now detected as red spots after spraying. The detection limit after bromination is about 0.5 µg tin per spot.

Waggon and Jehle[260,261] have reported on the quantitative detection of triphenyl and tri-, di-, and monobutyltin species in aqueous solution by a combination of liquid-liquid extraction, thin-layer chromatography and anodic stripping voltametry.

**Miscellaneous**

Glocking[191] has studied the degradation of organotin compounds in river waters.

### 6.3.2  SEAWATER

**Spectrofluorimetric method**

Fluorimetry has been used to determine triphenyltin compounds in seawater[247]. Triphenyltin compounds in water at concentrations of 0.004–2 pg $L^{-1}$ are readily extracted into toluene and can be determined by spectrofluorimetric measurements of the triphenyltin-3-hydroxyflavone complex.

**Hydride generation–gas chromatography–atomic absorption spectrometry**

Studies by Braman and Tompkins[245] have shown that non-volatile methyltin species $Me_nSn_{(4-n)}$ ($n = 1-3$) are ubiquitous at ng $L^{-1}$, concentrations in natural waters including both marine and freshwater sources. Their work, however, failed to establish whether tetramethyltin was present in natural waters because of the inability of the methods used to effectively trap this compound during the combined preconcentration purge and reductive derivatisation steps employed to generate volatile organotin hydrides necessary for tin specific detection.

Tin compounds are converted to the corresponding volatile hydride ($SnH_4$, $CH_3$, $SnH_3$, $(CH_3)_2SnH_2$, and $(CH_3)_3SnH$) by reaction with sodium borohydride at pH 6.5 followed by separation of the hydrides by gas chromatography and then detection by atomic absorption spectroscopy using a hydrogen-rich hydrogen–air flame emission type detector (Sn–H band). The apparatus used is shown in Figures 6.11 and 6.12.

The technique described has a detection limit of 0.01 ng as tin and hence parts per trillion of organotin species can be determined in water samples.

Braman and Tompkins[245] found that stannane ($SnH_4$) and methylstannanes ($CH_3SnH_3$, $(CH_3)_2SnH_2$ and $(CH_3)_3SnH$) could be separated well on a column comprising silicone oil OV-3 (20% w/w) supported Chromosorb W. A typical

# ANALYSIS OF ORGANOMETALLIC COMPOUNDS IN THE ENVIRONMENT

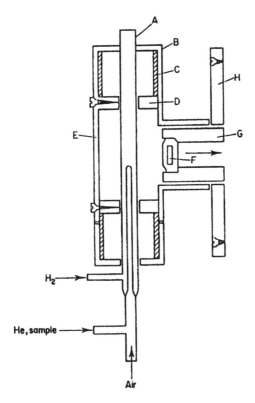

**Figure 6.11.** Quartz burner and housing. A, quartz burner; B, PVC cap; C, PVC tubing; D, mounting ring; E, PVC T-joining, 1.25 inch; F, filter and holder; G, PVC coupling; H, connection to photomultiplier housing (PM). (From Braman and Tompkins[245]. Reproduced by permission of American Chemical Society.)

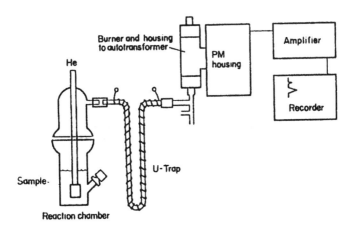

**Figure 6.12.** Appartus arrangement for tin analysis. (From Braman and Tompkins[245]. Reproduced by permission of American Chemical Society.)

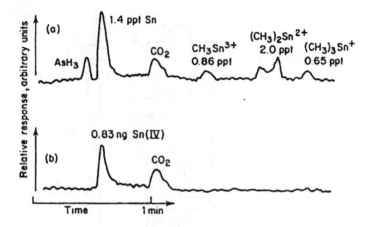

**Figure 6.13.** Environmental sample analysis and blank: separation of methyl stannanes (a) environmental analysis, Old Tampa Bay; (b) typical blank. (From Braman and Tompkins[245]. Reproduced by permission of American Chemical Society.)

separation achieved on a coastal water sample is shown in Figure 6.13. Average tin recoveries from seawater are in the range 96–109%.

A number of estuaral waters, from in and around Tampa Bay, Florida, area were analysed by this method for tin content. All samples were analysed without pretreatment. Samples which were not analysed immediately were frozen until analysis was possible. Polyethylene bottles, 500 mL volume, were used for sample acquisition and storage. The results of these analyses appear in Table 6.26, and the average total tin content of estuarine waters was 12 ng $L^{-1}$. Approximately 17–60% of the total tin present was found to be in methylated forms. This procedure, although valuable in itself, is incomplete in that any monobutyltin present escapes detection. Excellent recoveries of monobutyltin species are achieved with tropolone.

Jackson et al.[246] collected organotin compounds in seawater on Tenox GC. This column was then purged with helium onto a gas chromatographic column equipped with a flame photometric detector.

Valkirs et al.[263,264] compared two methods for the determination of $\mu$g $L^{-1}$ levels of dialkyltin and tributyltin species in marine and estuarine waters. The two methods studied were hydride generation followed by atomic absorption spectrometry and gas chromatography with flame photometric detection. Good agreement was obtained between the results of the two methods. Down to 0.01 mg $kg^{-1}$ of butyltin compounds, including tri-$n$-butyl tin and tri-$n$-butyl tin oxide, could be detected.

Studies on the effect of storing frozen samples prior to analysis showed that samples could be stored in polycarbonate containers at minus 20°C for 2–3 months without significant loss of tributyltin.

Brinckmann and co-workers[265] used a gas chromatographic method with or without hydride derivativisation for determining volatile organotin compounds

**Table 6.26.** Analysis of estuarine water samples*. (From Braman and Tompkins[245]. Reproduced by permission of American Chemical Society.)

| Sample | Tin(IV) ng L$^{-1}$ | % | Methyltin ng L$^{-1}$ | % | Dimethyltin ng L$^{-1}$ | % | Trimethyltin ng L$^{-1}$ | % | Total tin ng L$^{-1}$ |
|---|---|---|---|---|---|---|---|---|---|
| Sarasota Bay | 5.7 | 47 | 3.3 | 27 | 2.0 | 16 | 1.1 | 9.1 | 12 |
| Tampa Bay | 3.3 | 27 | 8.0 | 66 | 0.79 | 6.5 | n.d. | | 12 |
| McKay Bay | 20 | 88 | n.d. | | 2.2 | 9.6 | 0.45 | 2.0 | 23 |
| Hillsborough Bay | n.d. | | d.d. | | 1.8 | 71 | 0.71 | 29 | 2.5 |
| Hillsborough Bay, Seddon Channel North | 12 | 86 | 0.74 | 5.3 | 0.91 | 6.6 | 0.35 | 2.5 | 14 |
| Hillsborough Bay, Seddon Channel South | 13 | 83 | n.d. | | 2.4 | 15 | 0.31 | 1.9 | 16 |
| Manatee River | 4.8 | 61 | 1.4 | 1.7 | 1.1 | 14 | 0.65 | 8.2 | 7.9 |
| Alafia River | 3.4 | 73 | n.d. | | 0.75 | 16 | 0.55 | 12 | 4.7 |
| Palm River‡ | 567 | 98 | n.d. | | 4.6 | 0.80 | 4.0 | 0.69 | 576 |
| Bowes' Creek | 8.6 | 42 | 8.5 | 42 | 3.3 | 16 | n.d. | | 20 |
| Average | 7.9 | 63 | 2.4 | 19 | 1.7 | 14 | 0.46 | 3.7 | 12 |

*Data are average of duplicates.
†n.d. less than 0.01 ng L$^{-1}$ for methyltin compounds and 0.3 ng L$^{-1}$ for inorganic tin.
‡This set of values was not used in computing the average.

(e.g. tetramethyltin), in seawater. For non-volatile organotin compounds a direct liquid chromatographic method was used. This system employs a Tenax GC polymeric sorbent in the automatic purge and trap sampler coupled to a conventional glass column gas chromatograph equipped with a flame photometric detector. Flame conditions in the flame photometric detector were tuned to permit maximum respone to SnH emission in a H-rich plasma, as detected through narrow bandpass interference filters (610 ± 5 nm)[248]. Two modes of analysis were used: (1) volatile stannanes were trapped directly from sparged 10–50 mL water samples with no pretreatment; (2) volatilised tin species were trapped from the same or replicate water samples following rapid injection of aqueous excess sodium borohydride solution directed into the purge and trap vessel immediately prior to beginning the purge and trap cycle[246].

**Atomic absorption spectrometry**

Hodge et al.[238] determined butyltin chlorides in seawater by an atomic absorption spectrometric procedure.

**High performance liquid chromatography**

Ebdon and Alonso[266] determined tributyltin ions in estuarine waters by high performance liquid chromatography with fluorimetric detection using morin in a miscellar solution.

### Nuclear magnetic resonance spectroscopy

Laughlin et al.[267] analysed chloroform extracts of tributyltin dissolved in seawater using nuclear magnetic resonance spectroscopy. It was shown that an equilibrium mixture occurs which contains tributyltin chloride, tributyltin hydroxide, the aquo complex and a tributyltin carbonate species. Down to $0.4 \times 10^{-8} - 2 \times 10^{-8}$ µg $L^{-1}$ of organotin compounds could be determined.

### 6.3.3 RAINWATER

**Gas chromatography**

Braman and Tomkins[245] have developed methods for the determination of µg $L^{-1}$ amounts of inorganic tin and methyltin compounds in rain and other waters. Tin compounds are converted to the corresponding volatile hydride ($SnH_4$, $CH_3SnH_3$, $(CH_3)_2SnH_2$ and $(CH_3)_3SnH$) by reaction with sodium borohydride at pH 6.5 followed by gas chromatographic separation of the hydrides and then atomic absorption sepctroscopy using hydrogen rich hydrogen air flame emission type detector (Sn-H band).

The technique described has a detection limit of 0.01 ng as tin and hence parts per trillion or organotin species can be determined.

An average total tin content of rain was found to be 25 ng $L^{-1}$ and the methyltin form comprises 24% of that total. It is not clear why methyltin is predominantly found in rain waters but it may be a stable form resulting from demethylation of tetramethyltin.

### 6.3.4 SEDIMENTS

**Non-saline sediments**

**Gas chromatography**

Muller[253] has described a gas chromatographic method for the determination of tributyltin compounds in sediments. The tributyltin compounds are first converted to tributylmethyltin by reaction with ethyl magnesium bromide, and then analysed using capillary gas chromatography with flame photometric detection and gas chromatography-mass spectrometry. Tributyltin was found in samples of sediment and these results demonstrated that the technique has detection limits of less than 0.5 pg $L^{-1}$.

Lobinski et al.[250] speciated organotin compounds in sediment samples by capillary gas chromatography using helium microwave induced plasma emission spectrometry as a detector. They used the procedure to determine mono-, di-, tri- and some tetralkylated tin compounds in sediments. The ionic tin compounds were extracted as diethyldithiocarbamates into pentane then converted to pentyl magnesium bromide derivatives prior to gas chromatography. The absolute detection limit was 0.05 pg tin. Down to 13 ng $L^{-1}$ and 122 ng $L^{-1}$ respectively, of tributyltin and dibutyltin could be detected. Hattori[242] extracted alkyl and alkyltin compounds from sediments with methanolic hydrochloric acid and then, following mixture with sodium chloride and water, the mixture was extracted with benzene and converted to hydrides with sodium borohydride and analysed by gas chromatography using an electron capture detector. Down to 0.02 mg kg$^{-1}$ organotin compounds in sediments could be determined with a recovery of 70–95%.

### Atomic absorption spectrometry

Stephenson and Smith[269] used graphite furnace atomic absorption spectrometry to determine tributyltin in sediments. Recoveries from spiked samples ranged from 72% to 111%. The detection limit was 2.5 mg kg$^{-1}$ of sample.

### Liquid chromatography

Epler et al.[268] used laser enhanced ionisation as a selective detector for the liquid chromatographic determination of alkyltin compounds in sediments. The analysis was performed on a 1-butanol extract of the sediment.

### Supercritical fluid chromatography

Cal et al.[270] carried in situ derivativisation and supercritical fluid extraction for the simultaneous determination of butyltin and phenyltin compounds in sediments.

### Mass spectrometry

Methyltin compounds in sediments have been converted to tin hydrides with sodium borohydride prior to ion monitoring by mass spectrometry[271].

## Saline sediments

### Hydride formation–atomic absorption spectrometry

The procedure described by Hodge[238] has been applied to the determination of alkyltin compounds in marine sediments. A similar procedure has been applied by Randall et al.[272] to the determination of down to $0.6 \times 10^{-6}$ mg kg$^{-1}$ of methyl and butyltin compounds in estuary water sediments.

### Gas chromatography

Sinex et al.[273] determined methyltin compounds in amounts down to 3–5 pg (as Sn absolute), i.e. the sub µg kg$^{-1}$ level, in marine sediments by a procedure involving reaction with sodium borohydride to produce tin hydrides, followed by purge and trap analysis then gas chromatography with mass spectrometric detection.

## 6.3.5 SEWAGE SLUDGE

Muller[254] has described a procedure for the preconcentration and determination of mono, di, tri and tetra substituted organotin compounds in sewage sludge and lake water samples. The ionic compounds were extracted from diluted aqueous solutions as chlorides using a Tropolin-C18 silica cartridge and from sediments and sludges by using an ethereal tropolone solution. The extracted compounds were then ethylated by a Grignard reagent, and analysed by high-resolution gas chromatography with flame photometric detection.

## 6.3.6 BIOLOGICAL MATERIALS

### Spectrofluorimetric methods

Arakawa et al.[274] have shown that Morin (2', 3, 4', 5,7-pentahydroxyflavone) can be used as a fluorescence reagent for organotin, especially dialkyltin compounds. Although quercetin and 3-hydroxyflavone are similar to Morin in structure, they are unsuitable because of their sensitivity and instability. Morin produces a green fluorescence with various organotin compounds in organic solvent. The reagent is especially sensitive to dialkyltin compounds. The excitation and emission spectra show peaks at ca. 415 nm and ca. 495 nm, respectively, for each alkyltin-Morin complex and at ca. 405 nm and ca. 520 for the triphenyltin-Morin complex. The maximum fluorescence requires a ratio of 3 to 9 mol of Morin for 1 mol of dialkyl- and triphenyltin and 6–12, Mole of Morin for 1 mole of trialkyltin. Detection limits are $1 \times 10^{-9}$ M for dialkyltin, $1 \times 10^{-7}$ M for monoalkyltin. $5 \times 10^{-7}$ M for trialkyltin and $1 \times 10^{-7}$ M for triphenyltin. The fluorometric procedure can be used for the determination of individual organotin compounds in biological samples such as animal organs and urine following their prior separation by a suitable chromatographic technique. Recoveries of organotins added to various tissues at the 1.0–100 nmol level ranged from 91.0 to 99.7% depending upon the organotin species.

In this procedure, a $n$-hexane or ethyl acetate extract of a hydrochloric and homogentate of the biological sample was mixed with ethanolic Morin solution and examined fluorimetrically. The formation of the organotin-Morin complexes progressed very rapidly at room temperature and the fluorescence intensities remained constant for hours. Particularly, the dialkyltin complexes were stable over a number of hours. Dialklytin compounds produced a much stronger fluorescence than other organotin compounds with Morin. For 1 μM of each organotin compound the relative fluorescence intensity was 10.2 for $BuSnCl_3$, 42.5 for $Me_2SnCl_2$, 99.2 for $Et_2SnCl_2$, 99.4 for $Pr_2SnCl_2$, 51.7 for $Bu_2SnCl_2$, 2.2 for $Et_3SnCl$, 2.8 for $Pr_3SnCl$, 1.6 for $BuSnCl$, and 12.7 for $Ph_3SnCl$ at the same instrument setting. The concentration detection limits for dialkyltin compounds were in the 1–10 nM range, at which other organotin compounds could not be detected. This large difference in fluorescent intensities among different organotin–Morin complexes appears to be dependent on the valence state of the metal.

Organolead compounds such as di- and triethyllead and organosilane compounds such as di- and monomethylsilane did not interfere at 1 mM under the condition used for the determination of organotin. Other organometallic compounds such as dimethyl arsenide and methyl- and ethylmercury chlorides did not fluoresce at all. Although inorganic aluminium(III), zinc(II), tin(IV), magnesium(II) and cadmium(II) produced a strong fluorescence, and manganese(II), selenium(IV), and mercury(II) produced a very weak fluorescence with Morin in water solutions, these organometallic compounds of arsenic and mercury did not interfere at $1 \times 10^{-3}$ M under the condition of the organotin determination. Arsenic(III) or arsenic(V), lead(II), chromium(III) or chromium(VI), copper(II) and iron(II or III) did not fluoresce even in water solutions.

**Atomic absorption spectrometry**

Han and Weber[275] obtained nearly 100% recopvery of methyl and butyltin compounds from spiked oyster samples with a detection limit of 11–25 µg kg$^{-1}$ of oyster sample. The technique is based on hydride generation–atomic absorption spectrometry. Smith[276] applied graphite furnace atomic absorption spectrometry to the determination of tributyltin compounds in fish. Jones[277] applied hydride generation atomic absorption spectrometryc to the determination of down to 1.1–2.5 ng (absolute) of butyl tin compounds in oyster.

The graphite furnace atomic spectrometric procedure of Stephenson and Smith[269] described in Section 6.3.4 has also been applied to the determination of tributyltin in tissues.

**Gas chromatography**

A variety of seashells have been analysed for organotin compounds using the gas chromatographic procedure described by Braman and Tomkins[245] (see Section 6.3.1). The average total tin content of seashell and egg samples was between 0.001–0.002 mg kg$^{-1}$ and methylated tin compounds were detected ($Me_4Sn^{3+}$ and $Me_2Sn^{2+}$). Less than 0.01 µg kg$^{-1}$ $Me_3Sn^+$ was present.

Higher concentration of tin in the seashells relative to the water in which they were found would indicate the presence of bioaccumulation process.

Gas chromatographic methods have been described for the determination of tetraalkyl and trialkyl tins[279] in amounts down to $10^{-12}$ g (absolute) as trialkyltin chlorides in biological materals. Unfortunately, these methods were not easily applicable to the determination of the dialkyl homologues because of their absorption and decomposition during chromatography.

Short[276] and Sasaki et al.[278] both used an atomic absorption detector in conjunction with a gas chromatograph in the analysis of fish. Tri-n-butyltin and di-n-butyltin compounds have been determined[280] in fish as their hydride derivatives by reaction gas chromatography. The method involves the formation of volatile hydrides in a precolumn reactor packed with solid sodium borohydride in the injection port. The method is sensitive to 100 pg of tri-n-butyltin (as tin).

Gas chromatography with flame photometric detection using quartz surface induced luminescence has been used[190] to determine butyltin compounds in mussels with an absolute detection limit of 0.3 pg tin for tetrapropyltin or 2–3 pg tin for trimethylpentyltin, dimethyldipentyltin or methyltripentyltin. Between 0.04 and 0.09 µg g$^{-1}$ organic tin was found in mussel samples.

**High performance liquid chromatography**

High performance liquid chromatography–hydride generation direct current plasma emission spectrometric technique referred to previously[258] for the determination of organotin compounds in water, discussed in Section 6.3.1, has been applied to the analysis of clams and tuna fish.

6.3.7 CROPS AND PLANTS

**Atomic absorption spectrometry**

Hodge et al.[238] used hydride generation–atomic absorption spectrometry to determine organotin compounds in algae. Francois and Weber[282] used a similar technique to determine methyl and butyltin compounds in eel grass (*Zostera marina L*)

**Gas chromatography**

Gauer et al.[281] have described a gas chromatographic method for the determination of the residues of tricyclohexylhydroxystannane and its dicyclohexyl metabolite on strawberries, apples and grapes that have been treated with Pictran miticide. Crop samples were treated with aqueous hydrobromic acid to form bromoderivatives of the organotin compounds and these derivatives were extracted into benzene. When the residue levels were less than that 1 µg L$^{-1}$ the derivative solution was cleaned up on a column of silica gel. The derivatives were determined by gas chromatography at 200°C on a column packed with 2% of OV-225 on Chromosorb GAW-DMCS or at 100°C on a column packed with 0.5% of OV-225 on glass beads with helium as a carrier gas. Background interference was minimised by use of a halide-sensitive Coulson detector. Recovery of 1 mg L$^{-1}$ of added tricyclohexydroxystannane was 80–95%; that 0.1 mg l$^{-1}$ was 78 to 89%. Conditions are also described for the gas chromatographic determination of cyclohexylstannane acid, another possible degradation product of pictran.

**Chronopotentiometry**

Chronopotentiometry has been applied to the determination of triphenyltin acetate at very low concentrations in plant material. In this method a hanging-drop electrode is used at which the ions are reduced in a pre-electrolysis step at $-0.7$ V or a silver–silver chloride saturated potassium chloride electrode for 5 min, the potential is then increased gradually to $-0.1$ V, and the anodic diffusion current is registered at about $-0.45$ V. The sample for analysis is obtained by extraction of plant material with chloroform, the extract is washed with 0.1 N potassium hydroxide and 0.5 N potassium tartrate, then mineralised with sulphuric acid–nitric acid and the residue is dissolved in 5 N hydrochloric acid. A peak height of about one $\mu$A is obtained for a concentration of about 0.8 µg L$^{-1}$ tin in the plant digest.

## 6.4 ORGANOCADMIUM COMPOUNDS

Until 1996 organocadmium compounds had not been detected in the environment. Pongrats and Henman[358] using a differential pulse anodic scanning voltammetric method found low levels of methyl cadmium compounds in the Atlantic Ocean. Levels in the South Atlantic were approximately 700 pg L$^{-1}$, and those in the North Atlantic were below the detection limit of the method, i.e. below 470 pg L$^{-1}$. It is believed that these compounds were formed as a result of biomethylation of inorganic cadmium.

## 6.5 ORGANOARSENIC COMPOUNDS

### 6.5.1 NATURAL AND POTABLE WATERS

**Sample digestion procedures**

Most of the classical procedures for decomposing any organoarsenic compounds present in samples prior to the determination of total inorganic arsenic incorporate some mode of wet or dry digestion to destroy any organically bound arsenic, in addition to any other organic constituents present in the sample.

Probably the most frequently used method of digestion incorporates the use of nitric and sulfuric acids. Kopp[284] used this digestion method and experienced 91 to 114% recovery of arsenic trioxide added to deionised water and 86–100% recovery of the compound added to river water. The uncertaintities seem to arise when reviewing digestive methods using nitric and sulfuric acid. First, the addition of inorganic arsenic to an organic matrix and subsequent recovery of all the inorganic arsenic added is not definite proof of total recovery of any organoarsenicals present. Secondly, the choice of *o*-nitrobenzene arsenic acid and *o*-arsanilic acid by Kopp[284] in his recovery studies seems unfortunate since both compounds present arsenic attached to an aromatic ring which is a typical of cacodylic acid and disodium methyl arsonate, two widely used organoarsenicals.

Aside from nitric and sulfuric acid, a relatively simple digestive method employing 30% hydrogen peroxide in the presence of sulfuric acid was reported by Kolthoff and Belcher[285] and subsequently used by Dean and Rues[79] to determine arsenic in triphenylarsine.

Armstrong et al.[286] observed that organic matter in sea water could be oxidised to carbon dioxide on exposure to sufficient ultraviolet radiation from a medium pressure mercury arc vapour lamp. This approach has also been used to decompose organoarsenicals giving 111% recovery for o-arsenilic acid, 97% for sodium cacodylate and 108% for arsenazo.

## Spectrophotometric methods

Various workers have studied the spectrophotometric determination of organoarsenic compounds[284-292].

Stringer and Attrap[288] applied the Dean and Rues[79] sulfuric acid-hydrogen peroxide and the Armstrong et al.[286] ultraviolet decomposition methods to the determination of organoarsenic compounds in waste water.

Arsenic determinations were performed by either the silver diethyldithiocarbamate[287] or the arsine-atomic absorption method spectrophotometric procedure[284].

The organoarsenicals investigated were disodium methanearsonate, dimethylarsinic acid, and triphenylarsine oxide. All the digestive methods gave quantitative arsenic recoveries for the three organoarsenic compounds when added to wastewater samples. The ultraviolet photodecomposition proved to be an effective digestive technique, requiring a 4 h irradation to decompose a primary settled raw wastewater sample containing spiked quantities of the three organoarsenicals.

Between 96 and 105% recovery of triphenyl arsine oxide, disodium methanearsonate and dimethylarsinic acid at the 5 µg arsenic level were obtained by this procedure in water samples.

Haywood and Rile[290] have described procedures for the determination of arsenic in natural waters. Whilst this method does not include organic arsenic species, these can be rendered reactive either by photolysis with ultraviolet radiation or by oxidation with potassium permanganate or a mixture of nitric acid and sulfuric acids. Arsenic(V) can be determined separately from total inorganic arsenic after extracting arsenic(III) as its pyrrolidine dithiocarbamate into chloroform.

In the method for inorganic arsenic the sample is treated with sodium borohydride added at a controlled rate. The arsine evolved is absorbed in a solution of iodide and the resultant arsenate ion is determined spectrophotometrically by a molybdenum blue method. The detection limit is 0.14 µg $L^{-1}$ to 0.5 µg $L^{-1}$ depending upon the type of water. Silver and copper cause serious interference at concentrations of a few tens of mg $L^{-1}$; however, these elements can be removed either by preliminary extraction with a solution of dithizone in chloroform or by ion-exchange chromatography.

**Atomic absorption spectrometry**

Fishman and Spencer[293], Agemian and Chau[81], Stringer and Attrep[288] and Kotthoff and Belcher[285] used digestion with sulfuric acid–hydrogen peroxide or exposure to ultraviolet light to decompose organoarsenic compounds to inorganic arsenic prior to the determination of the latter by atomic absorption spectrometry (or spectrophotometry).

Stringer and Attrep[288] compared hydrogen peroxide–sulfuric acid digestion and ultraviolet light photodecomposition methods for the decomposition of three organoarsenic compounds in water samples (triphenylarsine oxide, disodium ethane arsonate and dimethylarsinic acid) to inorganic arsenic prior to reduction to arsine and determination by atomic absorption spectroscopy or by the silver diethyldithiocarbamate spectrophotometric method[287].

Stringer and Attrep[288] carried out their ultraviolet radiation photodecompositions using a medium pressure 450 W mercury arc photochemical lamp (Hannovia lamp no 67A0100) mounted vertically in the middle of an aluminium cylinder large enough in diameter to accommodate ten 1 inch diameter silica tubes. A blower fan was attached to the bottom of the lamp housing and served as an extractor fan to provide cooling of the lamp and samples. Excessive cooling of the lamp and resulting dimming of the lamp discharge was alleviated by enclosing the lamp in a 2 inch diameter quartz liner tapered at one end to retard air flow. The silica tubes which contained the samples were approximately 20 cm long and set 2 cm parallel to the lamp. Prior to irradiation, each sample was acidified with 3 drops of nitric acid and also 3 drops of 30% hydrogen peroxide. Following irradiation, each sample was transferred to a 100 mL volumetric flask and made to volume. Than, 10 mL of the 100 mL were transferred to a 125 mL Erlenmeyer flask, which also served as the arsine generator. The inorganic arsenic content was determined colorimetrically by atomic absorptions spectrophotometry, or, with silver diethyldithiocarbamate.

Kolthoff and Belcher[285] used a digestion procedure employing sulfuric acid and hydrogen peroxide for the decomposition of organoarsenic compounds prior to their determination in natural water by atomic absorption spectroscopy. This procedure was applied to three organoarsenic compounds of known purity.

Aliquots of each of the three organoarsenicals containing 5 μg of arsenic were added to 125 mL arsine generators. To each organic compound and a complete set of arsenic trioxide standards, 5 mL of concentrated sulphuric acid and 5 mL of 30% hydrogen peroxide were added. Samples and standards were boiled off and fumed for an additional 2 min. The samples were cooled and to each flask the following were added in succession: 25 mL of water, 7 mL of concentrated, hydrochloric acid, 5 mL of 15% potassium iodide and after 10 min, 5 drops of 40% stannous chloride. The samples were refrigerated for 20 min to allow for reduction to occur. Arsine was generated into an absorption tube containing 4 mL of silver diethyldithiocarbamate reagent for 1 hour then determined spectrophotometrically. Alternatively, following hydrogen peroxide–sulfuric acid decomposition inorganic arsenic was determined by atomic absorption spectroscopy.

The percentage recoveries obtained by the hydrogen peroxide–sulfuric acid digestion followed by spectrophotometric evaluation using silver diethyldithiocarbamate and by atomic absorption spectrometric determination are given in Table 6.27. The recoveries obtained by the same method when applied to arsenic spiked water samples are shown in Table 6.28.

Figure 6.14 shows the effect of ultraviolet irradiation as a function of time for triphenylarsine oxide, disodium methanearsonate and dimethylarsinic acid. The extent of arsenic recovery using photo-oxidation in conjunction with high sensitivity analysis when applied to water samples is shown in Table 6.29.

**Table 6.27.** Recovery of arsenic from triphenylarsine oxide, disodium methanearsonate, and dimethylarsinic acid employing wet digestion with 5 mL of 30% hydrogen peroxide and 5 mL of sulphuric acid with analyses by silver diethyldithiocarbamate and atomic absorption. (From Kolthoff and Belcher[285]. Reproduced by permission of John Wiley & Sons, New York.)

| Compound | Recovery (%)* | |
| --- | --- | --- |
| | AgDDC | Atomic absorption |
| Triphenylarsine oxide | 101.0 ± 5.2 (4) | 100.8(2) |
| Disodium methanearsonate | 103.5 ± 1.2 (4) | 100.8(2) |
| Dimethylarsinic acid | 99.9 ± 2.3 (3) | 99.2(2) |

*Each sample contained 5 µg arsenic. Number in parentheses indicates the number of samples run.

**Table 6.28.** Recovery of arsenic from triphenylarsine oxide, disodium methanearsonate, and dimethylarsinic acid spiked into water sample with wet digestion employing 15 mL of 30% hydrogen peroxide and 5 mL of sulfuric acid with analysis by silver diethyldithiocarbamate* (From Kolthoff, and Belcher[285]. Reproduced by permission of John Wiley & Sons, New York.)

| Sample | Recovery (%)† | |
| --- | --- | --- |
| | Sample 1 | Sample 2 |
| 100 mL AF | (9.7 µg As L$^{-1}$) | (6.9 µg As L$^{-1}$) |
| 100 mL AF + triphenylarsine oxide | 92.4 | 94.1 |
| 100 mL AF + disodium methanearsonate | 90.4 | 89.1 |
| 100 mL AF + dimethylarsinic acid | 96.0 | 90.2 |
| 100 mL SBE | (22.6 µg As L$^{-1}$) | (18.5 µg As L$^{-1}$) |
| 100 mL SBE + triphenylarsine oxide | 101.1 | 100.6 |
| 100 mL SBE + disodium methanearsonate | 98.1 | 104.4 |
| 100 mL SBE + dimethylarsinic acid | 89.4 | 96.6 |

*All 5 µg weights are as arsenic. †corrected for amount of arsenic initially present.

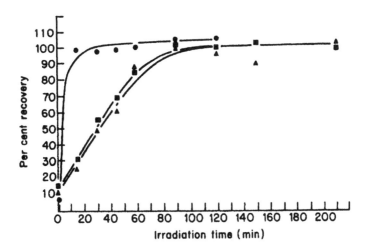

**Figure 6.14.** Recovery of arsenic after ultraviolet exposure as a function of time, ●, triphenylarsine oxide; ■ sodium arsonate; ▲ dimethylarsinic acid. (From Kolthoff and Belcher[285]. Reproduced by permission of John Wiley & Sons Ltd, Chulester.)

**Table 6.29.** Recovery of arsenic from triphenylarsine oxide, disodium methanearsonate, and dimethylarsinic acid spiked into water samples with digestion by u.v. and analysis by high sensitivity atomic absorption. (From Kolthoff, and Belcher [285]. Reproduced by permission John Wiley & Sons, New York.)

| Compound | Recovery (%)* | | |
| --- | --- | --- | --- |
| | Sample 1 | Sample 2 | Sample 3 |
| Triphenylarsine oxide | 110.0 | 100.0 | 102.8 |
| Disodium methanearsonate | 100.0 | 102.4 | 102.8 |
| Dimethylarsinic acid | 100.0 | 109.8 | 88.6 |

*Corrections were made for arsenic present in the samples.

The conclusion to be drawn from this work is that the digestive method employing hydrogen peroxide and sulfuric acid with analysis by either method gives arsenic recoveries ranging from 89.1 to 104.4% on natural water samples. This same digestive method, when applied to primary settled raw sewage, gave arsenic recoveries ranging from 89.1 to 96.0%. Arsenic recoveries of 89.4 −104.4% were experienced from an activated sludge effluent sample.

Regarding the ultraviolet photodecomposition procedure, Kolthoff and Belcher[285] showed that a 15 min exposure of triphenylarsine oxide resulted in greater than 99% photodecomposition of the compound. The monoalkylated arsenic compound reacted much more slowly than triphenylarsine oxide, requiring 2 h for complete decomposition so, obviously, this method has to be used with caution. A 4 hour decomposition produced 100–110% conversion of three organoarsenic compounds.

**Vapour generation gas chromatographic methods**

Various workers have converted organoarsenic compounds to organoarsines using sodium borohydride as a reducing agent, then determined the arsines by gas chromatography using various types of detectors[81,285,288,294–310]. The types of gas chromatographic detectors employed include electron capture or flame ionisation[195,297,303,305–309], atomic absorption spectrometric[288,299,303,305–309], spectrophotometric[288], emission spectrometric[296], mass specific detectors[310], multiple ion detectors[144], dc discharge microwave emission spectrometric[298] and neutron activation analysis[301].

In a sequential volatilisation varient of this procedure[296,299,300] applicable to samples which contain organoarsenic compounds the sample is gas purged to remove volatile organoarsines which are collected in a liquid nitrogen trap. The arsines are then separated by slow warming (sequential volatilisation) of the trap or are separated by gas chromatography and then measured by atomic absorption spectrometry or using electron capture or flame ionisation detectors.

Soderquist et al.[308] determined hydroxydimethyl arsine oxide in water by converting it to iododimethylarsine using hydrogen iodide followed by determination at 105°C on a column (15 ft × 0.125 in) packed with 10% of DC-200 on Gas Chrom Q (60–80 mesh) with nitrogen as carrier gas (20–30 mL min$^{-1}$) and electron capture detection. The recovery of hydroxydimethylarsine oxide (0.15 mg L$^{-1}$) added to pure water was 92.3% with a standard deviation of ±7.4%.

Edmond and Francesconi[295] have reported that the alkylated arsesicals methylarsonic acid and dimethylarinic acid occurring in the environment may be estimated directly by vapour generation atomic absorption spectrometry at 193.7 μm without prior digestion. Sodium borohydride treatment produces methylarsine, and dimethylarsine, respectively, which are swept directly into a hydrogen–nitrogen entrained air flame by the excess hydrogen generated by hydrolysis of the sodium borohydride. These methylated arsines are estimated in a manner identical to the arsine produced following acid digestion or dry ashing. The calibration curves and instrument responses are directly comparable and are dependent only on the quantity of arsenic entering the flame.

The sample containing sodium arsenate, methylarsonic acid, and dimethylarsinic acid was treated with sodium borohydride and the mixed arsines generated trapped in a glass bead packed tube (200 mm × 25 mm) at −180°C. The cooling agent was removed and the trap allowed to warm slowly in the laboratory atmosphere. A valve assembly kept the trap sealed and was released periodically with a simultaneous flow of nitrogen through the trap into the flame. The valve was opened for 5 s each min. The instrument responses was recorded on a chart moving at 2 mm min and resembled, in outline, a gas chromatographic trace with arsine peaking at 3 min methylarsine at 8 min and dimethylarsine at 13 min.

Detection limits were 500 ng for inorganic arsenic and 1 μg for monomethylarsonic acid and dimethylarsinic acid. Aqueous solutions of methylarsonic acid and dimethylarsinic acid were adjusted to 2% in hydrochloric acid before addition of the sodium borohydride solution.

Arsenic itself, originating by the reduction of inorganic arsenic and methyl arsines, originating by the reduction of organoarsenic compounds, have different responses in colorimetric versus atomic absorption methods of analysis. The colorimetric diethyl-dithiocarbamate method is much more responsive to arsine than to methyl arsines, whilst vapour generation atomic absorption spectrometry is equally responsive to both forms of arsenic, depending only on the weight of arsenic entering the flame.

Lussi-Schlatter and Brandenberger[310] have reported a method for inorganic arsenic and phenylarsenic compounds based upon gas chromatography with mass specific detection after hydride generation with head space sampling. However, methylarsenic species were not examined.

Parris et al.[309] have studied in detail the chemical and physical considerations that apply in the determination of trimethylarsine using an atomic absorption spectrophotometer with a heated graphite tube furnace as a detector for a gas chromatograph. 5 μg arsenic could be detected by this technique.

Gifford and Bruckenstein[297] generated the gaseous hydrides of AsIII, SnII and SbIII by sodium borohydride reduction and separated them on a column of Poropak Q. Detection was at a gold gas porous electrode by measurement of the respective electro-oxidation potentials. Detection limits were (5 mL samples) AsIII 0.21 μg L$^{-1}$, SnII 0.8 μg L$^{-1}$ and SbIII 0.2 μg L$^{-1}$.

Odanaka et al.[144] have reported that the combination of gas chromatography with multiple ion detection system and a hydride generation heptane cold trap technique is useful for the quantitative determination of arsine, monomethyl-, dimethyl-, and trimethylarsenic compounds and this approach is applicable to the analysis of environmental and biological samples.

In this method, arsine and methylarsines produced by sodium borohydride reduction are collected in $n$-heptane ($-80°C$) and then determined. The limit of detection for a 50 mL sample was 0.2–0.4 μg L$^{-1}$ of arsenic. Relative standard deviations ranged from 2% to 5% for distilled water replicates spiked at the 10 μg L$^{-1}$ level. Recoveries of all four arsenic species from river water ranged from 85% to 100%.

Dimethyl arsinic acid yields predominantly dimethylarsine, whilst methylarsonic acid yields predominently methylarsine and trimethylarsine oxide yields predominently trimethylarsine. In this method aqueous samples (1–50 mL) that had been previously neutralised with hydrochloric acid and/or sodium bicarbonate were placed in the reaction vessel and were diluted to 60 mL with water. Then 6 mL of 4 N hydrochloric acid and 2 mL of methanol were added. The cold trapping system was then connected and the carrier gas (helium) was allowed to pass through the system at 100–150 mL min$^{-1}$ for 1 min to flush out any air. The reduction was initiated by injecting 3 mL of 10% sodium borohydride solution through the rubber septum into the aqueous sample. The volatile arsines were collected in a $n$-heptane (3–5 mL) cold trap for 2 min. The low temperature was maintained by submerging the trap in a dry ice-acetone ($-80°C$) bath. The helium flow was continued for 1 min to ensure complete generation and trapping of the arsines. After the collection of the arsines, the $n$-heptane (5 mL) was injected into the gas chromatograph/mass spectrometer.

The following ions were characteristic and intense ions in the mass spectra of arsines, arsine $m/z$ 78 M$^+$, 76 (M $-$ 2)$^+$, methylarsine; $m/z$ 92 M$^+$, 90 (M $-$ 2)$^+$, 76 ((M $-$ CH$_3$) $-$ 1)$^+$, dimethylarsine $m/z$ 106 M$^+$, 90 ((M $-$ CH$_3$) $-$ 1)$^+$, trimethylarsine, $m/z$ 120 M$^+$, 105 ((M $-$ CH$_3$) $-$ 1)$^+$, 103 ((M $-$ CH$_3$)$^+$ $-$ 2)$^+$. To achieve simultaneous measurement and to assess the specificity of the analysis, for instance the $m/z$ 76, 78, 89 and 90 were monitored for arsine, methylarsine, and dimethylarsine and/or $m/z$ 90, 103, 105 and 106 for alkylarsines as methylarsine, dimethylarsine and trimethylarsine. However, simultaneous determination of all four arsenicals could not be done at one injection because of a limited range of detectable mass spectra in the system used.

Andreae[299] described a sequential volatilisation method for the sequential determination of arsenate, arsenite, mono-, di, di- and trimethylarsine, monomethylarsonic and dimethylarsinic acid, and trimethylarsine oxide in natural waters with detection limits of several ng L$^{-1}$. The arsines are volatilised from the sample by gas stripping; the other species are then selectively reduced to the corresponding arsines and volatilised. The arsines are collected in a cold trap cooled with liquid nitrogen. They are then separated by slow warming of the trap or by gas chromatography and measured with atomic absorption, electron capture and/or flame ionisation detectors. He found that these four arsenic species all occurred in natural water samples.

The apparatus for the volatilisation and trapping of the arsines is constructed from Pyrex glass, with Teflon stopcocks and tubing and with Nylon Swagelok connectors. The sample trap consists of a 6 mm o.d. Pyrex U-tube of $ca.$ 15 cm length, filled with silane treated glass wool. The interior parts of the six-way valve which interfaces the volatilisation system with the gas chromatograph are made of Teflon and stainless steel.

The gas chromatograph is equipped with a $^{63}$Ni electron capture detector mounted in parallel with a flame ionisation detector and an auxiliary vent by the use of a column effluent splitter. The separation is performed on a 4.8 mm o.d., 6 m long stainless steel column packed with 16.5% silicone oil DC-550 on Chromosorb W AW DMCS. The helium carrier gas flow rate is 80 mL min$^{-1}$.

The electron capture detector (Hewlett-Packard 2-6195 with a $^{63}$Ni source) is operated in the constant pulse mode with a pulse interval of 50 μs. The atomic absorption detection system consists of a Varian AA5 with a hollow cathode arsenic lamp; the standard burner head is replaced by a 9 mm i.d. quartz burner cuvette.

To isolate the arsine species the sample 1–50 mL is introduced into the gas stripper with a hypodermic syringe through the injection port.

Any volatile arsines in the sample are stripped out by bubbling a helium stream through the sample. Then 1 mL of the Tris buffer solution for each 50 mL sample is added, giving an initial pH of about 6. Into this solution 1.2 mL of 4% sodium borohydride solution is injected while continuously stripping with helium. After about 6–10 min the As(III) is converted to arsine and stripped from the solution. The pH at the end of this period is about 8. Then 2 mL of 6 N hydrochloric acid is added, which brings the pH to about 1. The addition of three aliquots of 2 mL of 4% sodium borohydride solution during 10 min reduces As (IV), monomethylarsonic acid, dimethylarsinic acid, and trimethylarsine oxide to the corresponding arsines, which are swept out of the solution by the helium stream coming from the reaction vessel stream.

In Table 6.30 are reported some values obtained for inorganic arsenic, monomethyl arsonic acid (MMAA) and dimethylarsinic acid (DMAA) in natural water samples.

Andreae[299] commented that, if stored in airtight containers, the free arsines are stable in solutions for a few days. They are slowly oxidised by traces of air to the corresponding acids. From untreated samples, the methylated acids are lost measureably after a period of about three days, depending on the initial concentrations. They are stable indefinitely if the sample is made 0.05 N in hydrochloric acid. Arsenite is slowly oxidised to arsenate in samples below 0.05 µg L$^{-1}$ As(III), a loss of arsenite become detectable after about one week. Acidification of the sample increases the oxidation rate and arsenic loss can be detected after one day. If the samples are stored in a freezer below $-15°C$ or

**Table 6.30.** Arsenic species concentrations in natural waters, (µg L$^{-1}$ As). (From Andreae[299]. Reproduced by premission of American Chemical Society.)

| Locality and Sample Type | As(III) | As(V) | MMAA | DMAA |
|---|---|---|---|---|
| Seawater, Scripps Pier, La Jolla, CA | | | | |
| 5 Nov. 1976 | 0.019 | 1.75 | 0.017 | 0.12 |
| 11 Nov. 1976 | 0.034 | 1.70 | 0.019 | 0.12 |
| Seawater, San Diego Trough Surface | 0.017 | 1.49 | 0.005 | 0.21 |
| 25 m below surface | 0.016 | 1.32 | 0.003 | 0.14 |
| 50 m below surface | 0.016 | 1.67 | 0.003 | 0.004 |
| 75 m below surface | 0.021 | 1.52 | 0.004 | 0.002 |
| 100 m below surface | 0.060 | 1.59 | 0.003 | 0.002 |
| Sacramento River, Red Bluff, CA | 0.040 | 1.08 | 0.021 | 0.004 |
| Owens River, Bishop, CA | 0.085 | 42.5 | 0.062 | 0.22 |
| Colorado River, Parker, AZ | 0.114 | 1.95 | 0.063 | 0.051 |
| Colorado River, Slough Near Topcock CA | 0.085 | 2.25 | 0.13 | 0.31 |
| Saddleback Lake, CA | 0.053 | 0.020 | 0.002 | 0.006 |
| Rain, La Jolla, CA | | | | |
| 10 Sept. 1976 | 0.002 | 0.180 | 0.002 | 0.024 |
| 11 Sept. 1976 | 0.002 | 0.094 | 0.002 | 0.002 |

under dry ice, an initial loss of arsenite corresponding to about 0.02 µg L$^{-1}$ is obtained. The sample then remains unchanged with prolonged storage.

## High performance liquid chromatography

Jyn Yynn[311] chelated organoarsenic compounds with sodium bis (trifluoroethyl) dithiocarbamate prior to application of high performance liquid chromatography. He applied the technique to mixtures of arsenite, arsenate, dimethylarsonic acid, dimethylarsinic acid and other organoarsenic compounds.

Blais et al.[315] determined arsenobetaine, arsenochline and tetramethyl arsonium ion in non-saline waters in amounts, respectively, down to 13.3, 14.5 and 7.8 µg by a procedure based on liquid chromatography–thermochemical hydride generation atomic absorption spectrometry.

## Ion-exchange chromatography

Ion-exchange chromatography has been used to achieve separations of monomethyl-arsonate, dimethylarsinic and tri- and pentavalet arsenic. Dietz and Perez[308] have described methods which separated the inorganic arsenic from each of the organic species using ion exchange chromatography. Here, further inorganic speciation relies on redox-based colorimetry (Johnson and Pilson[313]). Both the accuracy and precision suffer from the low As(III)/As(IV) and As(total)/P ratios normally encountered in the environment. Henry and Thorpe[314] determined these four arsenicals by coupling a digestion and reduction scheme with ion-exchange chromatography. However, the utility of this technique for routine environmental analysis is limited, since the implementation time is substantial. This method also relies on estimating As(V) by difference.

Grabinski[316] has described an ion-exchange method for the complete separation of the above four arsenic species, on a single column containing both cation and anion exchange resins. Flameless atomic absorption spectrometry with a deuterium arc background correction is used as a detection system for this procedure. This detection system was chosen because of its linear response and lack of specificity for these compounds combined with its resistance to matrix bias in this type of analysis.

The elution sequence was as follows: 0.006 M trichloroacetic acid (pH 2.5), yielding first As(III) and then monomethylarsonate; 0.2 M trichloroacetic acid yielding As(V): 1.5 M NH$_4$OH followed by 0.2 M trichloroacetic acid yielding dimethylarsinite. Detection was by flameless atomic absorption spectrometry. Arsenic recoveries (full-procedure) ranged from 97% to 104% for typical lake water samples; more erratic but still acceptable recoveries (98% to 107%) were obtained from arsenic contaminated sediment interstitial water. The overall analytical detection limit was 10 µg L$^{-1}$ (original sample mixture) for each individual arsenic species. Relative standard deviations ranged from 0.7% to

1.3%. respectively for lake water and distilled deionised water replicates spiked at the 500 µg L$^{-1}$ level.

Grabinski[316] spiked 0.500 µg of each of the four arsenic species into filtered (0.45 µM) lake water and distilled and deionised water. Arsenic recoveries for the entire procedure averaged 104%, 100% and 97% and 99% for As(III), monomethylarsonate, As(IV) and dimethylarsinite, respectively. Relative standard deviations for replicate determinations ranged from 0.7% for As(III) and dimethylarsinite to 1.3% for As(V).

Aggett and Kadwani[317] report the development and application of a relatively simple anion-exchange method for the speciation of arsenate, arsenite, monomethylarsonic acid and dimethylarsinic acid. As these four arsenic species are weak acids the dissociation constants of which are quite different it seemed that separation by anion-exchange chromatography was both logical and possible.

Aggett and Kadwani[317] employed a two stage single column anion-exchange method using sodium hydrogen carbonate and chloride as eluate anions. These species appear to have no adverse effects in subsequent analytical procedures. Its successful application is dependent on careful control of pH. Analyses were performed by hydride generation atomic absorption spectroscopy.

Separation of arsenic(III) and dimethylarsinic acid by elution with sodium hydrogen carbonate was satisfactory in the pH range 5.2–6.0. Elution of monomethylarsonic acid was accelerated and satisfactory separation from arsenic(V) achieved using saturated aqueous carbon dioxide (pH 4.0–4.2) containing 10 g L$^{-1}$ of ammonium chloride. In order to avoid oxidation of trivalent to pentavalent arsenic on the column, it was necessary to pretreat the resin with 1 M nitric acid and 0.1 M EDTA before use.

The application of these methods to interstitial waters and lake sediment and also species of lake weed revealed no methylated arsenic species, only As(III) and As(IV) in the 0.5 µg L$^{-1}$ range were found.

**Differential pulse polarography**

Yamamoto[322] and Dietz and Perez [312] observed that dimethyl arsinite has a strong affinity for acid-charged cation exchange resins. Elton and Geiger[318] used this fact to separate monomethyl arsonate and dimethyl arsinite prior to determinations of the organoarsenicals by differential pulse polarography. The authors reported detection limits of 100 µg L$^{-1}$ and 300 µg L$^{-1}$ respectively for these two compounds. Henry et al.[319] reported a method for the determination of arsenic(III) and arsenic(V) and total inorganic arenic by differential pulse polarography. Arsenic(III) was measured indirectly in 1 M perchloric acid or 1 M hydrochloric acid[320]. Total inorganic arsenic was determined in either of these supporting electrolytes after the reduction of electroinactive arsenic(V) to arsenic(III) with aqueous sulphur dioxide. Arsenic(V) was evaluated by difference. Sulphur dioxide was selected because it reduced arsenic(V) rapidly and quantitatively, and excess reagent was readily removed from the reaction mixture.

## 6.5.2 SEA WATER

### Atomic absorption spectrometry

Persson and Irgum[321] determined sub ppm levels of dimethyl arsinate by preconcentrating the organoarsenic compound on a strong cation-exchange resin (Dowex AG 50 W-XB). By optimising the elution parameters, dimethyl arsinate can be separated from other arsenicals and sample components, such as group I and II metals, which can interfere in the final determination. Graphite-furnace atomic absorption spectrometry was used as a sensitive and specific detector for arsenic. The described technique allows dimethyl arsinate to be determined in a sample (20 mL) containing a $10^5$ fold excess of inorganic arsenic with a detection limit of 0.02 ng As per mL. Good recoveries were obtained from the artificial seawaters, even at the 0.05 µg $L^{-1}$ level, but for natural seawater samples the recoveries were lower (74–85%). This effect could be attributed to organic sample components that eluted from the column together with dimethyl arsinate.

Various workers[295,299,323,325–328] have described procedures for the determination of dimethyl arsinate, monomethyl arsinite and trimethyl arsine oxide in seawater based on conversion of the organoarsenic compounds to hydrides using sodium borohydride, cold trapping the hydrides, then vaporising them by controlled warming of the cold trap and atomic absorption spectrometry of the vapourised fractions.

Haywood and Riley[290] determined organoarsenic compounds in seawater in amounts down to 0.14 µg $L^{-1}$ by a procedure based on pholysis with ultraviolet light, extraction of organic arsenic compounds with dipyrrolidine dithiocarbamate in chloroform then atomic absorption spectrometry of the extract.

### Gas chromatography

Talmi and Bostick[298] extracted methylarsenic compounds from seawater with cold toluene, then analysed the extract by gas chromatography using a mass spectrometric detector. Down to 0.25 mg $L^{-1}$ of organoarsenic compounds were detectable.

### Column chromatography

Yamamoto[322] separated organoarsenic compounds from seawater by column chromatography. The organoarsenic compounds were reduced to arsine with sodium borohydride and analysed by atomic absorption spectrometry.

## 6.5.3 SEDIMENTS AND SOILS

### Gas chromatography

Odanaka et al.[144] have reported the application of gas chromatography with multiple ion detection after hydride generation with sodium borohydride to the determination of mono-, di-, methyl arsenic compounds, trimethylarsenic oxide and inorganic arsenic in soil and sediments; this work is discussed more fully in Section 6.5.1 (i.e. organoarsenic compounds in water). Recoveries in spiking experiments were 100–102% (mono and dimethylarsenic compounds and inorganic arsenic) and 72% (trimethyl arsenic oxide).

Maher[331] has described a method for the determination of down to 0.01 mg kg$^{-1}$ of organoarsenic compounds in marine sediments. In this procedure the organoarsenic compounds are separated from an extract of the sediment by ion exchange chromatography, the isolated organoarsenic compounds are reduced to arsines with sodium borohydride and collected in a cold trap. Controlled evaporation of the arsine fractions and detection by atomic absorption spectrometry completes the analysis.

### 6.5.4 INDUSTRIAL EFFLUENTS AND WASTE WATERS

**Spectrophotometric method**

Sandhu and Nelson[332] have also studied the interference effects of several metals on the determination of organically bound arsenic at the 0–100 µg L$^{-1}$ range in waste water by the silver diethyldithiocarbamate method. Antimony and mercury interfere specifically, forming complexes with silver diethyldithiocarbamate at absorbance maxima at 510 and 425 nm respectively. Recovery of organic arsenic released by digesting solutions was tested and shown to be about 90%.

**Differential pulse polarography**

Henry et al.[314,319] separated monomethyl arsonate, dimethylarsinite, arsenic(III) and arsenic(IV) on a Dowex 50 W-8X cation (H$^+$ form) and AGI-X8 cation exchange column from samples of pond water receiving fly ash slurry from a coal fired power station. They then determined these substances by differential pulse polarography. The above four arsenic species are present in natural water systems. However, a dynamic relationship exists whereby oxidation–reduction and biological methylation–dimethylation reactions[302,333–335] provide pathways for the interconversions of the arsenicals. Analytical methods capable of distinguishing between the predominant species of arsenic are necessary if immediate and potential impacts are to be accurately assessed. This method is described below.

Cation, Dowex 50W-X8 and anion, AG1-X8 exchange resins. To remove potential interference, wash the 50–100 mesh resins with alternating 0.5 M solutions of hydrochloric acid and sodium hydroxide at flow rates of 5–10 mL min$^{-1}$. Slurry pack the cation exchange resin, in the H$^+$ form unto a 2.0 cm i.d. glass column to a height of 16 cm. After use, regenerate the column by a reaction with 1.0 M hydrochloric acid. Slurry pack the anion exchange resin into an 8 cm i.d. column to a height of 11 cm. Convert the resin to the acetate form by passing 250 mL of 0.5 M sodium hydroxide over the column followed by

100 mL of 1.0 M sodium acetate, both at flow rates of 5–10 mL min$^{-1}$. Rinse the column with triply distilled water then a mobile phase that was 0.1 M in total acetate concentration and had a pH of 4.7 until the pH of the effluent was also 4.7. Figure 6.15 depicts the analytical procedure for separation of the subsequent determination of each of the arsenicals. The pH of the sample is adjusted to between 4 and 10, and the sample divided into four aliquots. Two of these are used to determine arsenic(III) and total inorganic arsenic as described by Henry et al.[319] The concentration of arsenic(V) is calculated from the difference between these two values. The polarograph used by Henry et al.[319] was the Princeton Applied Research Corporation Model 174A analyser, using a dropping mercury electrode (0.845 mg S$^{-1}$, drop time 2.0551). A third of 200 mL aliquot is mixed with 2.0 mL of 1.74 M acetic acid and passed through the cation exchange resin at a flow rate of 5 mL min$^{-1}$ to isolate the dimethylarsinite from the matrix. The sample is followed by a mobile phase consisting of 0.02 M acetic acid. Arsenic(III) and arsenic(V) and monomethylarsinate elute within 70 mL. Dimethylarsinate is recovered by stripping the column with 1.0 M sodium hydroxide at 1.0 mL min$^{-1}$. Eluate coeluted from 31–42 mL after addition of the basic mobile phase contains the dimethylarsinate.

The fraction of eluate containing dimethylarsinic acid is added to 7 mL of 70% perchloric acid contained in a flask fitted with an air condenser. The resulting solution is heated to fumes and heated at 200°C for 2.5 h. The solution is cooled and diluted with 10 mL of triply distilled water. The arsenic(V) produced in this

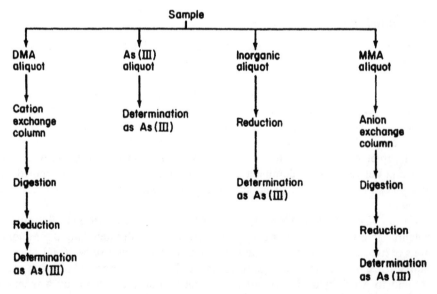

**Figure 6.15.** Flow chart for the determination of As(III), As(V), monomethylarsinic acid (MMA) and dimethylarsinic acid (CMA). (From Henry and Thorpe[314]. Reproduced by permission of American Chemical Society.)

digestion is reduced to arsenic(III) and determined according to the procedure of Henry and Thope[314].

Monomethylarsonate is isolated using the anion exchange column. A 50 mL aliquot of the sample is mixed with sodium acetate–acetic acid buffer to obtain a total acetate concentration of 0.01 M and a pH of 4.7. The resulting solution is loaded onto the column and this is followed with 50 mL of a mobile phase of similar composition. Arsenic(III) and dimethylarsinite will elute within the first 100 mL (sample + eluant). Monomethylarsonate is retained on the resin. The flow rate is constant (1–2 mL min$^{-1}$) throughout the procedure.

To resolve monomethylarsonate from arsenic(V) the total concentration of the mobile phase is increased to 0.1 M while maintaining a constant pH. Eluate containing monomethylarsonate is collected between 20 and 40 mL from the introduction of the more concentrated phase. Arsenic(V) begins to appear after 75 mL from the same reference point.

The fraction containing monomethylarsonate in perchloric acid is digested under the same conditions as is dimethylarsinite. The reaction proceeds at a faster rate however, and reaches completion within 30 min. Determine the arsenic(V) produced in this reaction by the procedure of Henry et al.[319]

Henry and Thorpe[314] showed that digestion of dimethylarsinite decomposed this substance at the boiling point of the perchloric acid water azeotrope (203°C) in a minimum of 2.5 h with a recovery of 92–107%. The efficiency of the perchloric acid digestion and the recovery of dimethylarsinite from the entire procedure was in the range of 95–108%. The mean recoveries and standard deviations indicate the procedures are quantitative and reproducible. The detection limit for dimethylarsinite is 8 μg L$^{-1}$. The recovery of monomethyl arsonate was 90–107% and the detection limit 18 μg L$^{-1}$.

Table 6.31 shows results obtained in determinations of the four arsenic species in a spiked water sample. The recoveries of the known additions of the four arsenic species substantiate the accuracy of the procedure and these results demonstrate the utility of the method for the speciation of arsenicals in complex aqueous matrices.

**Table 6.31.** Results of the analysis of an Environmental Protection Agency standard reference sample*. (From Henry and Thorpe[314]. Reproduced by permission of American Chemical Society.)

| Species | Concentration present (μg L$^{-1}$) | Determination | | | | Average concentration | Standard deviation |
|---|---|---|---|---|---|---|---|
| | | 1 | 2 | 3 | 4 | | |
| As(III) | 26 | 25 | 33 | 26 | 34 | 30 | 3.7 |
| As(V) | 82 | 85 | 81 | 76 | 76 | 80 | 4.4 |
| MMA | 83 | 72 | 86 | 85 | 79 | 80 | 6.5 |
| DMA | 69 | 69 | 71 | 63 | 65 | 67 | 3.7 |

*Values in ppb arsenic, MMA, monomethylarsonate acid; DMA, dimethylarsinic acid. Reprinted.

## 6.5.5 BIOLOGICAL MATERIALS

**Atomic absorption spectrometry**

Maher[331] and Agemian and Cheam[81] determined mono- and dimethyl arsenic compounds in fish in amounts down to less than 1 mg kg$^{-1}$ by procedures based on controlled reduction to arsines by sodium borohydride. Reduction is followed by cold trapping of arsines, controlled fractionation and detection by atomic absorption spectrometry. Arsenobetaine has been determined in crab by a procedure based on inductively coupled plasma atomic emission spectrometry[336].

Beauchemin et al.[338] identified and determined organoarsenic species in dogfish muscle reference samples using high performance liquid chromatography coupled to an inductively coupled plasma mass spectrometer as a detector. They also applied thin-layer chromatography, electron impact mass spectrometry and graphite furnace atomic absorption spectrometry. About 84% of the total arsenic present in the dogfish was in the form of arsenobetaine (16 µg Agg$^{-1}$). The absolute detection limit of the hplc–icpaes technique was 0.3 µg as arsenic.

**Gas chromatography**

Odanaka et al.[144] have reported the application of gas chromatography with multiple ion detection after hydride generation with sodium borohydride to the determination of mono, dimethyl arsenic compounds, trimethylarsenic oxide and inorganic arsenic in animal tissues; this work is discussed more fully in Section 6.5.1. Recoveries in spiking experiments were 87–103%.

This work showed that in living animal tissue administered inorganic arsenic may be biomethylated to monomethylated, dimethylated and even trimethylated compounds.

Schwedt and Russe[337] have described a method for the gas chromatographic determination of arsenic (as triphenylarsine) in biological material. In this method, the dry sample is burnt in a Schoniger oxygen combustion flask containing 3 N hydrochloric acid, the products are washed out with 3 N hydrochloric acid and aqueous potassium iodide and aqueous sodium bisulphite are added to the solution which is then extracted with diethyldithiocarbamate solution in dichloromethane. The extract is evaporated and the residue is stirred for 30 min with diphenylmagnesium solution in ethyl ether. After addition of dilute sulfuric acid the separated ether phase is evaporated and the residue is treated with mercaptoacetic acid solution. After being set aside for 20 min, the solution is chromatographed on a glass column (2 × 3 mm) packed with 5% of teraphthalic acid-treated Carbowax 20 M on Gas-Chrom Q (80–100 mesh) and operated at 220°C with nitrogen, helium or argon as carrier gas and flame ionisation detection. Down to 2 mg L$^{-1}$ of arsenic in the sample could be determined by this procedure.

Odanaka et al.[262] identified dimethylarsinate (salt of dimethylarsinic acid $(CH_3)_2$ $AsO_2H$ DMAA) as a metabolite in the blood, urine and faeces of rats which had been administered orally with doses of ferric methane arsenate $(CH_3AsO_3)_3Fe_2$; MAF). Following separation of inorganic arsenic and methane–arsonate by cellulose thin-layer chromatography, dimethylarsinate was converted to dimethylarsine $((CH_3)_2AsH_2)$ by reduction and identified by gas chromatography–mass spectrometry.

**Ion-exchange chromatography**

Ion chromatography can separate organoarsenic compounds in the liquid phase. Ion-exchange resins have been employed for the separation of As(III), As(V), methylarsonic acid and dimethylarsinic acid in biological samples. Stockton and Irgolic[339], separated As(III), As(V) arsenobetaine, and arsenocholine by high-performance liquid chromatography using a reversed phase ion suppression technique. Ion-exchange high performance liquid chromatography seemed applicable to the separation of these arsenicals with better resolution than the conventional ion-exchange chromatography using resin.

Because a selective and sensitive detection is necessary after the separation, atomic absorption spectrometry has been used for this purpose[295]. DC plasma and microwave helium plasma atomic emission spectrometry have been employed for gas-phase detection[299,312]. For the liquid phase, a graphite furnace Zeeman effect atomic absorption method has been used with the automated sampler. Inductively coupled argon plasma emission spectrometry seems another choice as a high performance liquid chromatography detector because it has high sensitivity for arsenic, low chemical interference and wide dynamic range.

Maher[331] used ion-exchange chromatography to separate inorganic arsenic and methylated arsenic species in marine organisms and sediments. The method determines monomethylarsenic, and dimethylarsenic. The procedure involves the use of solvent extraction to isolate the arsenic species which are then separated by ion-exchange chromatography and determined by arsine generation.

Hanumura et al.[135] applied thermal vaporisation and plasma emission spectrometry to the determination of organoarsenic compounds in fish.

### 6.5.6 PLANTS AND CROPS

**Spectrophotometric method**

White[340] determined organoarsenic compounds in marine brown algae by a procedure based on generation of arsine and quantification by the silver diethyldithiocarbamate spectrophotometric method.

**Atomic absorption spectrometry**

The sodium borohydride reduction-atomic absorption spectrophotometric method[331] described in Sections 6.4.3 and 6.5.5 for the determination of organoarsenic compounds in sediments and fish has also been applied to the determination of monomethyl arsenic acid and dimethylarsinic acid in macro algae.

### Gas chromatography

Odanaka et al.[144] have reported the application of gas chromatography with multiple ion detection after hydride generation with sodium borohydride to the determination of mono and dimethylarsenic compounds, trimethylarsenic oxide and inorganic arsenic in plant materials; this work is discussed more fully in Section 6.5.1 (organoarsenic compounds in water). Recoveries in spiking experiments were 92–103%.

## 6.6 ORGANOANTIMONY COMPOUNDS

### 6.6.1 NATURAL WATERS

#### Atomic absorption spectrometry

Andraea et al.[259] have described a method for the determination of methylantimony species, Sb(III) and Sb(V) in natural waters by atomic absorption spectrometry with hydride generation. The limit of detection was 0.3–0.6 µg $L^{-1}$ for a 100 mL sample. Some results are also reported for estuary and seawaters.

Talmi and Norwel[298] have studied the application of the gas chromatography–microwave plasma detector technique to the analysis of organoantimony compounds in environmental samples[144].

#### Polarography

Besada and Ibrahim[341] have described a polarographic method for the determination of antibilharzial organoantimony compounds in river waters.

## 6.7 ORGANOSELENIUM COMPOUNDS

### 6.7.1 NATURAL WATERS

#### High performance liquid chromatography

Killa and Robenstein[342] determined selenols, diselenides and selenyl sulphides in environmental samples by reverse phase liquid chromatography with electrochemical detection. Selenols were detected at the downstream electrode of an electrochemical detector with dual mercury/gold amalgam electrodes. Diselenides and selenyl sulphides were reduced at the upstream electrode to selenols or thiols, which were then detected at the downstream electrode.

#### Isotope dilution mass spectrometry

Tanzer and Hennman[343] used this technique to determine dissolved selenium species including selenate, selenite and trimethylselenonium in environmental water samples with a detection limit of 200 pg $g^{-1}$ to 15 ng $g^{-1}$.

## 6.7.2 BIOLOGICAL MATERIALS

**Neutron activation analysis**

Trimethyl selenonium, selenite and total seleno amino acids in biological fluids, including urine and serum have been separated[344] by anion exchange chronatography and determined by neutron activation analysis. The detection limits for the determination of trimethyl selenium/selenite and selenoamino acids were 10 and 40 µg L$^{-1}$ as selenium, respectively.

## 6.8 ORGANOGERMANIUM COMPOUNDS

### 6.8.1 NATURAL AND ORGANIC WATERS

**Atomic emission spectrometry**

Braman and Tompkins[196] have described an atomic emission spectrometric method for the determination of inorganic germanium and methylgermanium (and inorganic antimony) in amounts down to 0.4 ng in environmental samples. These compounds are reduced to hydrides using sodium borohydride, then separated prior to atomic emission spectrography.

**Hydride generation atomic absorption spectrometry**

Hambrick et al.[195] observed germanium compounds in some natural waters which are reduced and trapped similarly to $Ge(OH)_4$, but which elute from chromatographic packings after $GeH_4$. They suggested that these peaks are unidentified methylgermanium species by analogy with previous observations for arsenic, antimony and tin. Further work confirmed the presence of methylgermanium in natural waters and led them to modify their technique in order to optimise the recovery of the methylgermanium species. The technique used by Hambrick et al.[195] was a modification of the method developed earlier for inorganic germanium by Andreae and Froelich[345].

Inorganic and methylgermanium species were determined in aqueous matrix at the parts-per-trillion level by a combination of hydride generation, graphite furnace atomisation, and atomic absorption spectrometry. The germanium species were reduced by sodium borohydride to the corresponding gaseous germanes and methylgermanium hydrides, stripped from solution by a helium gas stream, and collected in a liquid nitrogen-cooled trap. The germanes were released by rapid heating of the trap and enter a modified graphite furnace at 2700°C. The atomic absorption peak was recorded and electronically integrated. The absolute detection limits are 155 pg of germanium for inorganic germanium 120 pg of germanium for monomethylgermanium, 175 pg of germanium for dimethylgermanium and 75 pg of germanium for trimethylgermanium. The precision of the determination ranges from 6% to 16%. The method was applied to marine, freshwater and rain water samples. The major germanium species in sea water is monomethylgermanium. Trimethylgermanium was found in sea water. Total germanium values for sea water are slightly lower than those reported by Braman and Tompkins[245] (79 ng L$^{-1}$).

Jin et al.[189] determined germanium species in natural water and waste water by hydride generation inductively coupled argon plasma–mass spectrometry. Absolute detection limits were as follows: inorganic germanium (0.08 pg), methyl germanium trichloride and dimethyl germanium dichloride (0.1 pg) and trimethyl germanium chloride (0.09 pg).

**Solvent extraction of organogermanium compounds**

The following compounds are quantitatively extracted from hydrochloric acid solutions of water samples with carbon tetrachloride[346]; triethylgermanium chloride, trimethylgermanium chloride, diethylgermanium dichloride, dimethylgermanium dichloride, phenylgermanium trichloride, ethylgermanium trichloride, methylgermanium trichloride and germanium tetrachloride.

**Liquid–liquid extraction**

Methylated and inorganic germanium compounds have been separated from water by liquid–liquid extraction with organic liquids containing respectively charged oxygen donors[347].

## 6.9 ORGANOMANGANESE COMPOUNDS

### 6.9.1 NATURAL WATERS

**Gas chromatography**

Aue et al.[348] used gas chromatography with electron capture detection to determine methylcyclopentadienyl manganese tricarbonyl in natural waters.

**High performance liquid chromagraphy**

Walton et al.[184] separated organomanganese and organotin compounds by high performance liquid chromatography using laser excited atomic fluorescence in a flame as a high sensitivity detector.

The absolute detection limit for manganese was 8–22 pg for various organomanganese compounds including methyl cyclopentadienyl manganese tricarbonyl

## 6.9.2 AIR

**Gas chromatography-atomic absorption spectrometry**

Coe et al.[183] have described a gas chromatographic furnace–atomic absorption technique for the determination of methylcyclopentadienylmanganese at the µg m$^{-3}$ level in air samples. The method involves trapping of methylmanganese in a small segment of gas chromatographic column and then determination by gas chromatography with an electrothermal atomic absorption detector. The detection limit of the procedure is 0.05 ng m$^{-3}$. The method involves trapping of methylmanganese in a small segment of gas chromatographic column and then determination by gas chromatography with an electrothermal atomic absorption detector. A Pye gas chromatograph was interfaced to the graphite furnace using a tantalum connector. A glass chromatogaphic column (2.3 m long, 6 mm o.d.) was packed with 3% OV-1 on high performance Chromosorb W (80–100 mesh). The gas from the chromatograph was transferred to the furnace through Teflon-lined aluminium tubing. Samples of air were collected on Teflon-lined aluminium U-tubes (30 cm long, 3 mm o.d.) packed with 3% OV-1 on Chromosorb W (80–100 mesh). These tubes were placed in a water-ice cooling bath. Air entered through an air filter and was pumped through the U-tube at about 70 mL min$^{-1}$ using a vacuum pump. The length of sampling time and the average flow rate, checked frequently during sampling, were used to compute sample volume.

The gas chromatographic method referred to in Section 6.9.1[348] has also been applied to the determination of methyl cyclopentadienyl manganese tricarbonyl in air.

## 6.10 ORGANOCOPPER COMPOUNDS

### 6.10.1 NATURAL WATERS

**High performance liquid chromatography**

Brown et al.[349] have studied the application of reversed phase high performance liquid chromatography with molecular and atomic absorption detection to the separation and quantification of organocopper speciation in soil-pore waters. Polar dissolved organic compounds and associated copper complexes are separated using either a single Hypersil GDS column or two Hypersil ODS columns and a Hamilton PRP 1 column in series. Quantifications was achieved using ultraviolet detectors for the organic molecular species and graphite furnace atomic-absorption spectrometry for the copper. As well as the high relative molecular mass compounds, such as polysaccharides, peptides, lipids and humic substances, there is a wide range of low relative molecular mass metabolites produced in soils by micro-organisms and plant roots. Reversed phase columns will retain polar compounds most effectively when their ionization is suppressed. As many of the polar organic ligands present in pore waters are acidic (e.g. citric acid), the eluent system chosen to suppress ionisation was also acidic (0.02% V/V orthophosphoric acid, pH 2.6). A less acidic eluent was utilised for less polar compounds. Ammonium formate solution (0.01 M, pH 6.1) was found to be suitable because it gave a low signal to noise ratio for graphite furnace a.a.s. analysis. This eluent was used for substances with higher retention values than the citric acid-copper complex.

One soil-pore water examined contained five recognisable polar dissolved organocopper compounds as revealed by absorbance measurement at 215 nm on the eluate. Analysis of soil-pore waters using the single-column high performance liquid chromatographic system interfaced with graphite furnace atomic absorption spectroscopic system showed that the copper was not always associated with the polar dissolved organic compounds in the same proportions. In the majority of pore waters association of the copper with citric acid and neighbouring eluting compounds was found.

## 6.11 ORGANONICKEL COMPOUNDS

### 6.11.1 AIR

Nickel tetracarbonyl is one of the most dangerous chemicals known. It exhibits acute toxicity, carcinogenicity, tetraogenicity and can be produced spontaneously in unsuspecting environments whenever carbon monoxide contacts an active form of nickel.

**Fourier transform infrared spectroscopy**

Infrared spectrometry, Fourier transform IR (FTIR) chromatography and chemiluminescence have all been used to analyse for nickel carbonyl. All of these methods have the advantage of direct measurement with detection limits of about 1 µg $L^{-1}$ and are also adaptable to process stream monitoring. The methods based on the FTIR and the plasma chromatography methods have been compared in real sample analyses and agree within a few percent. The chemiluminescent analysis for nickel carbonyl demonstrates a detection limit of 0.01 µg $L^{-1}$ with a linearity over four orders of magnitude.

**Chemical ionisation mass spectrometry**

Campana and Risby[350] have studied the possibility of the formation of nickel carbonyl by the reaction of carbon monoxide (present in exhaust gases in the range 1–15%) with nickel containing catalysts (Monel) under typical automotive operating conditions.

Chemical ionisation mass spectrometry was used to determine trace levels of nickel carbonyl in a typical exhaust gas mixture. Also a catalytic flow reactor system was designed to study the formation of nickel carbonyl where the reacted gas was analysed for nickel carbonyl. This reactor approximated the conditions found in an automotive exhaust catalytic converter system. These workers found various positive ions which can be attributed to nickel carbonyl and its fragments in this mass spectrum. The effect of pressure on the intensity of both the reactant gas ions and the nickel carbonyl was studied in an attempt to find the optimum condition for the quantification of nickel carbonyl. Using these data, the minimum detectable limit was found to be 10 µg $L^{-1}$ for nickel carbonyl. This methodology was used to monitor nickel carbonyl in the effluent from a model reactor for a catalytic controlled automobile.

## 6.12 ORGANOSILICON COMPOUNDS

### 6.12.1 NATURAL WATERS

**Gas chromatography**

Mahone et al.[351] have described a method for the quantitative characterisation of water borne organosilicon compounds. Substances such as silanol or silanol-functional materials are converted to trimethylsilylated derivatives which can be determined by gasliquid chromatography. The method gives good accuracy and precision in the mg $L^{-1}$ range and with suitable precautions, can be extended to the low µg $L^{-1}$ range.

**Inductively coupled plasma atomic absorption spectrometry**

Wanatabe et al.[115] have described a simple and rapid method for the separation and determination of down to 10 µg $L^{-1}$ siloxanes in river water, using inductively coupled plasma emission spectrometry. Organosilicon compounds are extracted with petroleum ether then evaporated to dryness. The damp residue is dissolved in methyl isobutyl ketone and aspirated into the plasma.

Figure 6.16 shows a typical calibration curve obtained in this procedure. Concentrations of toluene, vegetable oil and mineral oil present at concentrations up to 10 times that of the siloxane did not interfere in this procedure. Recoveries of siloxane obtained by spiking river water samples were generally in the region of 70% with a coefficient of variation of about 5%.

**Mass spectrometry**

Van der Post[352] has described a method for the determination of silanols in water based on their ability to reduce nitrite or nitrate to ammonia at normal temperature. Individual silanols are identified by mass spectrometry.

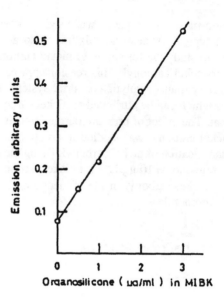

**Figure 6.16.** Calibration curve of organosilicone at the selected operating conditions. (From Wanatabe et al.[115]. Reproduced by permission of Elsevier Science Publishers BV, Netherlands.)

**Miscellaneous**

Glocking[191] studied the degradation of organosilicon compounds in river water.

## 6.12.2 BIOLOGICAL MATERIALS

**Spectroscopic methods**

Horner et al.[353] describe methods for the visible and infrared spectroscopic determination of trace amounts of silicones in biological materials. In the spectrophotometric method the sample is decomposed by wet oxidation and the residues are fused with sodium carbonate. The metals are dissolved in hydrochloric acid and the silicon is determined by the molybdenum blue procedure.

## 6.13 ORGANOBORON COMPOUNDS

### 6.13.1 TISSUES

Barth et al.[354] used this technique to determine down to 0.1 mg kg$^{-1}$ of organoboron compounds (($Et_3NH)_2B_{12}H_{12}$, $CS_2B_{12}H_{11}SHH_2O$), $C_{15}H_{32}B_{10}O_5$ in tissues.

## 6.14 ORGANOPHOSPHOROUS COMPOUNDS

### 6.14.1 ENVIRONMENTAL SAMPLES

Kalesar and Walsar[355,356] detected organophosphorous compounds in environmental samples with a supported copper plus cuprous oxide island film using ac and dc. Organophosphorous compounds were detected at the μg $L^{-1}$ or mg $kg^{-1}$ level.

Grate et al.[357] developed a smart sensor system for the determination of traces of organophosphorous compounds in air. They detected vapours using a temperature controlled array of surface acoustic wave sensors (SAWS) with automated sample preconcentration and pattern recognition. Down to 0.01 μg $m^{-3}$ of organophosphorous compounds could be detected in air.

## 6.15 ORGANOSULFUR COMPOUNDS

The method described by Grate et al.[357] referred to above has also been applied to the determination of down to 0.5 mg $m^{-3}$ of organosulfur compounds in air.

## 6.16 SELECTION OF APPROPRIATE ANALYTICAL METHODS

The analytical method to be used will depend on the type of organometallic compound, its concentration range in the sample, the detection limit required and the type of sample. This information, taken from Sections 6.1–6.11, is summarised in Table 6.32.

Firstly, it must be explained that not all published procedures give details of the detection limits achievable. Where this data is available it is quoted in Table 6.33 and compared to the range of concentrations of organometallic compounds actually encountered in environmental samples (taken from Table 6.32).

This table highlights some, but not all, of the available methods that have sufficiently low detection limits to render them appliable to even the lowest concentrations of organometallic compounds encountered in various types of environmental samples. Thus, Table 6.33 is a useful starting point for the selection for any particular type of any organometallic compound and sample type of an appropriate analytical method that will cover the range of concentrations likely to be encountered in the sample including very low concentrations. The situation will depend on the type of sample, *viz* waters, sediments or living creatures. Therefore, these are discussed separately below.

### 6.16.1 WATER SAMPLES

These comprise river, potable, oceanic and coastal waters, also rain and snow.

**Table 6.32.** Concentration ranges of organometallic compounds encountered in environmental samples.

| Type of water | Type of organometallic compound | Range encountered (total compounds) in actual samples | Lowest concentration (individual compounds) quoted in reported analysis | Units |
|---|---|---|---|---|
| | **WATERS** | | | |
| | *Alkyl Mercury* | | | |
| River and Potable | Methylmercury | 0.005–1.1 | 0.005 (0.0003 potable) | µg L$^{-1}$ |
| Coastal | Methylmercury | 0.005–0.06 | 0.005 | µg L$^{-1}$ |
| Sea | Methylmercury | 0.005–0.06 | 0.005 | µg L$^{-1}$ |
| Rain | Alkylmercury | 0.00001–0.009 | 0.00001 | µg L$^{-1}$ |
| | *Alkyl Lead* | | | |
| River and Potable | Alkyl lead | 50–530 | 10 | µg L$^{-1}$ |
| Coastal | Alkyl lead | | | |
| Sea | Alkyl lead | | | |
| Rain | Alkyl lead | | | |
| | *Alkyl Tin* | | | |
| River and Potable | Methyltin | 0.002–0.05 | 0.005 | µg L$^{-1}$ |
| | (butyltin) | 0.05–473 | 0.005 | µg L$^{-1}$ |
| Coastal | Alkyltin | 0.00001–3.5 | 0.00001 | µg L$^{-1}$ |
| Sea | Methyltin | 0.00001–2.0 | 0.00001 | µg L$^{-1}$ |
| | butyltin | 0.00001–3.5 | 0.00001 | µg L$^{-1}$ |
| Rain | methyltin | 0.0005–0.006 | 0.0005 | µg L$^{-1}$ |
| | *Alkyl Arsenic* | | | |
| River and Potable | Alkylarsenic | | 0.001 | µg L$^{-1}$ |
| Coastal | Alkylarsenic | 0.001–2.6 | 0.001 | µg L$^{-1}$ |
| Sea | Alkylarsenic | 0.001–2.6 | 0.001 | µg L$^{-1}$ |
| Rain | — | — | — | |
| | **SEDIMENTS** | | | |
| | *Alkylmercury* | | | |
| River | Alkylmercury | 0.00001–0.004 | 0.00001 | mg kg$^{-1}$ |
| Coastal | Alkylmercury | 0.00001–0.0004 | 0.00001 | mg kg$^{-1}$ |
| | *Alkyl Lead* | | | |
| River | Alkyl lead | | 0.00001 | mg kg$^{-1}$ |
| Coastal | Alkyl lead | | 0.00001 | mg kg$^{-1}$ |
| | *Alkyl Tin* | | | |
| River | Butyltin | 0.05–0.3 | 0.01 | mg kg$^{-1}$ |
| Coastal | Alkyltin | 0.00001–0.08 | 0.00001 | mg kg$^{-1}$ |
| | *Alkyl Arsenic* | | | |
| River | Alkylarsenic | | 0.00001 | mg kg$^{-1}$ |
| Coastal | Alkylarsenic | | 0.00001 | mg kg$^{-1}$ |

Table 6.32. (continued)

| Type of water | Type of organometallic compound | Range encountered (total compounds) in actual samples | Lowest concentration (individual compounds) quoted in reported analysis | Units |
|---|---|---|---|---|
| **FISH AND INVERTEBRATES** | | | | |
| | Alkyl Mercury | | | |
| Fish | Methylmercury | 0.01–12 | 0.01 | mg kg$^{-1}$ |
| | Alkyl Lead | | | |
| Fish | Alkyl lead | 5–20 | 0.0001 | mg kg$^{-1}$ |
| Crustacea | Alkyl Lead | 0.01–0.05 | 0.00001 | mg kg$^{-1}$ |
| | Alkyl Tin | | | |
| Invertebrates | Alkyltin | 0.03–0.1 | 0.0005 | mg kg$^{-1}$ |
| | Butyltin | 0.02–0.1 | 0.0005 | mg kg$^{-1}$ |
| | Alkykl Arsenic | | | |
| Mollusc | Alkylarsenic | 0.00001–23 | 0.00001 | mg kg$^{-1}$ |
| Prawn | Alkylarsenic | 1.4–23 | 0.00001 | mg kg$^{-1}$ |
| **WATER PLANTLIFE** | | | | |
| | Alkyl Lead | | | |
| Phytoplankton | Alkyl lead | 0.04–22 | 0.04 | mg kg$^{-1}$ |
| | Alkyl Arsenic | | | |
| Algae | Alkylarsenic | 0.2–90 | 0.00001 | mg kg$^{-1}$ |
| Phytoplankton | Alkylarsenic | 0 | 0.00001 | mg kg$^{-1}$ |

**Alkyl mercury compounds**

The concentration ranges of these encountered in sea and river water samples are, respectively, 0.005–0.06 and 0.005–1.1 µg L$^{-1}$. Methods based on atomic absorption spectrometry with detection limits as low as 0.02 µg L$^{-1}$ [81,85] are adequate for the analysis of these types of sample although they are not sufficiently sensitive to determine alkyl mercury at the level encountered in rain (0.00001–0.009 µg L$^{-1}$).

**Alkyl lead compounds**

Ionic alkyl lead compounds in snow and presumably water can be determined by capillary gas chromatography using an atomic absorption spectrometric detector[211]. This method has a detection limit of 0.0000015 µg L$^{-1}$ well below the levels encountered in environmental waters (viz 0.0005–473 µg L$^{-1}$).

**Akyl and aryl tin compounds**

The levels of these encountered in environmental waters range from 0.00001–20 µg L$^{-1}$ (butyltin) to 0.00001–3.5 µg L$^{-1}$ (methyltin). A wide

**Table 6.33.** Choice of methods.

(A) *Environmental Waters and Snow*

| Type of water | Type of organo-metallic compound | Method | LD of the method ($\mu g\ L^{-1}$) | Reference | Conc. range found in environmental samples ($\mu g\ L^{-1}$) |
|---|---|---|---|---|---|
| Sea/River | MeHg | Benzene extraction—aas | 20 | 85 | 0.005–0.06 (coastal & sea) |
| Sea/River | MeHg | Cold vapour AAS | 0.02 | 81 | 0.005–1.1 (river & potable) |
|  |  |  |  |  | 0.00001–0.009 (rain) |
| Snow | $Et_3Pb^+$ $Et_2Pb^{2+}$ | Capillary glc—aas | 0.000001 | 211 | 10–530 (river & potable) |
| River water | $Et_4Sn$ $Me_4Sn$ | Laser ionisation glc—mass spectrometry | 0.0000015 (1 mL sample) | 255 | 0.002–0.05 (methtin) (river) |
|  |  |  |  |  | 0.05–473 (butyltin) (river) |
| River water | $Bu_4Sn$ $Ph_4Sn$ | Supercritical fluid Chromatography—inductively coupled plasma mass spectometry | 0.00004 (1 mL sample) | 257 | 0.0005–0.006 (methyltin) (rain) |
| Sea | $Bu_3Sn$ | Chloroform extraction—NMR | 0.000002 | 267 | 0.00001–20 (butyltin) (coastal and sea) |
|  |  |  |  |  | 0.00001–3.5 (methyltin) (coastal & sea) |
| Sea | $Ph_3Sn$ | Toluene extraction spectrophotometric | 0.000004 | 235 |  |
| Sea | $Bu_3Sn$ $Bu_2Sn$ $BuSn$ | Capillary glc—microwaved induced plasma emission spectrometry | 0.00005 (1 mL sample) | 250 |  |
| Sea | $BuSnCl$ | aas | 0.04 (1 mL sample) | 238 |  |
| Sea | $Bu_3Sn$ bis tributyl tin oxide | Methylisobutyl ketone extraction—graphite furnace aas also glc mass spectrometric or hydride reduction flame aas | 0.00001 | 263 264 |  |
| Sea | $MeSn$ $Me_2Sn$ $Me_3Sn$ | Reduction to $MeAsH_3$ $Me_2SnH_2$ and $Me_3SnH$—glc—aas | 0.001 (10 mL sample) | 245 |  |

**Table 6.33.** (*continued*)

| Type of water | Type of organo-metallic compound | Method | LD of the method ($\mu g\ L^{-1}$) | Reference | Conc. range found in environ-mental samples ($\mu g\ L^{-1}$) |
|---|---|---|---|---|---|
| River water | Arsenobetaine Arsenocholine Tetramethyl arsonium chloride | hplc—hydride generation—aas | 1 (10 mL sample) | 315 311 | 0.001 - (rivers) |
| Sea | MeAs | Toluene extraction—glc mass spectrometric detector | 0.25 | 298 | 0.001–2.6 (coastal and sea) |
| Sea | dimethyl arsinate | Preconcentration on Dowex AG5-graphite furnace aas | 0.02 | 321 | |
| Sea | Alkylarsenic | UV photolysis, chloroformic dithiocarbamate extraction—aas | 0.14 | 329 | |
| River water | MeGeCL$_3$ Me$_2$GeCl$_2$ Me$_3$GeCl | Hydride generation icpaes | 0.00001 (10 mL sample) | 189 | |

| Type of sediment | Type of organo-metallic compound | Method | LD ($mg\ kg^{-1}$) | Reference | Conc. range found in environ-mental samples ($mg\ kg^{-1}$) |
|---|---|---|---|---|---|
| *(B) Sediments* | | | | | |
| River and sea | MeHg | Cold vapour aas | 0.0025 | 112,108, 107,109 | 0.00001–0.004 (river) |
| River | MeHg | Steam distill-cold vapour aas | 0.0025 | 110 | 0.000 01–0.0004 (coastal & sea) |
| | MeHg | Cold vapour aas | 0.0002 | 113,218, 215 | |
| River | Alkyl Pb | Conversion to tetra alkyl Pb glc—aas detector | 0.015 | 205 | 0.000 01 - (rivers) 0.000 01 - (coastal & sea) |

Table 6.33. (continued)

| Type of sediment | Type of organo-metallic compound | Method | LD (mg kg$^{-1}$) | Reference | Conc. range found in environmental samples (mg kg$^{-1}$) |
|---|---|---|---|---|---|
| River and sea | MeSn | Reduction to hydride with NaBH$_4$, purge and cold trap controlled evaporation, glc with mass spectrometric detection | 0.000003–0.000005 | 273 | 0.05–0.3 (rivers) 0.00001–0.08 (coastal) |
| River and sea | MeSn BuSn | Ditto, detection by aas | 0.0000006 | 272 | |
| River | MeSn | Sodium borohydride reduction, ion monitoring by mass spectrometry | 0.000003–0.000005 (1 g sample) | 271 | |
| | Bu$_3$Sn Bu$_2$Sn BuSn | Capillary glc-helium microwave induced plasma emission spectrometric detection | 0.000005 | 250 | |
| River | Alkyl and Aryl tin | Benzene extraction conversion to hydride with sodium borohydride—glc with electron capture detector | 0.02 | 242 | |

| Type of creature | Type of organo-metallic compound | Method | LD (mg kg$^{-1}$) | Reference | Conc. range found in environmental samples (mg kg$^{-1}$) |
|---|---|---|---|---|---|
| (C) *Fish and Invertebrates* | | | | | |
| Fish | Alkyl Hg Alkyl Hg Aryl Hg | Atomic absorption spectrometry conversion to chloro derivatives—glc | 0.0005 0.001 | 40 118 | 0.01–12 (fish) |
| Fish | MeHg | Derivativization—glc with aas detector | 0.000004 (0.1 g sample) | 152 | |
| Fish | MeHg | hplc with aas detector | 0.037 (0.1 g sample) | 150 | |
| Fish | MeHg | hplc with electrochemical detector | 0.002 | 153 | |

Table 6.33. (continued)

| Type of creature | Type of organo-metallic compound | Method | LD (mg kg$^{-1}$) | Reference | Conc. range found in environ-mental samples (mg kg$^{-1}$) |
|---|---|---|---|---|---|
| Mussel | MeHg | Solvent extraction—glc | 0.01 | 74 | |
| Crustacea | Alkyl Hg | Solvent extraction—glc | 0.001 | 70 | |
| Shrimp, oyster clam, tuna | MeHgCl | glc with electron capture detector | 0.02<br>0.25 | 145<br>127,148, 149 | |
| Inverte-brates | Alkyl Hg | Ethylation with Na B(C$_2$H$_5$)$_4$, cryogenic trapping, glc with electron capture detector | 0.000004 | 152 | |
| Fish | Methyl ethylPb | glc with atomic absorption spectrometric detector | 0.025 | 201,202 | 0.01–0.05 (crustacea)<br>5–20 (fish) |
| Oyster | BuSn | Sodium borohydride reduction—aas | 0.011 (0.1 g sample)<br>0.0011 (1 g sample) | 277 | 0.03–0.1 (alkyltin) (invertebrates) |
| Oyster | MeSn<br>BuSn | Sodium borohydride reduction—aas | 0.000023(MeSn)<br>0.000025(BuSn) | 275 | 0.02–0.1 (butyltin) (invertebrates) |
| Mussel | Me$_3$SnC$_5$H$_{11}$<br>Me$_2$Sn(C$_5$H$_{11}$)$_2$<br>MeSn(C$_5$H$_{11}$)$_3$ | glc with flame photometric detector | 0.00003 (0.01 g sample)<br>0.000003 (0.1 g sample) | 190 | |

Table 6.33. (continued)

| Type of creature | Type of organo-metallic compound | Method | LD (mg kg$^{-1}$) | Reference | Conc. range found in environmental samples (mg kg$^{-1}$) |
|---|---|---|---|---|---|
| Fish | MeAs<br>Me$_2$As | Sodium borohydride reduction—aas | 0.01 | 331,81 | 0.00001–23 (mollusc) |
| Fish | Arsenobetaine | hplc with inductively coupled plasma mass spectrometric detector | 0.002 | 338 | 1.4–23 (prawn) |
| Oyster | Methylethyl lead | Solvent extraction glc—aas detector | 0.01 | 215,218 | |

glc—gas chromatography
aas—atomic absorption spectrometry
hplc—high performance liquid chromatography
NMR—Nuclear magnetic resonance spectroscopy
icpaes—Inductively coupled plasma atomic emission spectrometry.

variety of methods is available for analysing at these concentration ranges including supercritical fluid chromatography[257], NMR[267], capillary GLC[250], spectrometry[235], graphite furnace atomic absorption spectrometry[263,264]. These methods have detection limits down to 0.000 002 µg L$^{-1}$.

**Organoarsenic compounds**

These have been found in sea and coastal water in the range 0.001–2.6 µg L$^{-1}$. Gas chromatographic[118,298] and atomic absorption spectrometric[329] have, respectively, detection limits of 0.25 and 0.14 µg L$^{-1}$.

The most sensitive method available involving preconcentration on Dowex AG50 followed by atomic absorption spectrometry[321] extends the detection limit down to 0.02 µg L$^{-1}$ fairly near to the lowest organoarsenic content of 0.001 µg L$^{-1}$ encountered in marine waters.

6.16.2 SEDIMENTS

**Methylmercury compounds**

These have been encountered in river and marine waters, respectively in the concentration ranges 0.000 01–0.004 mg kg$^{-1}$ and 0.000 01–0.0004 mg kg$^{-1}$. Available methods involving atomic absorption[107−110,112], with a detection limit of 0.0002 mg kg$^{-1}$ are insufficiently sensitive to analyse at the 0.000 01 mg kg$^{-1}$ level and it would seem that a preconcentration technique would be needed to effect analysis of these very low levels.

**Alkyl lead compounds**

The same comments apply as discussed under alkylmercury compounds.

**Alkyltin compounds**

Alkyltin compounds have been encountered in river waters and coastal water sediments in the concentration ranges, respectively, of 0.05–0.3 mg kg$^{-1}$ and 0.000 01–0.08 mg kg$^{-1}$. A variety of methods is available for analysis at even the lowest end of these concentration ranges including reduction to tin hydrides followed by gas chromatography[272] or mass spectrometry[271] or an alternative method involving capillary column gas chromatography with a helium microwave induced plasma emission spectrometric detector[250]. These methods having detection limits down to 0.000 0006 mg kg$^{-1}$, i.e. are more than adequate for handling the sample concentration ranges quoted above.

### 6.16.3 FISH AND INVERTEBRATES

**Alkylmercury compounds**

Alkylmercury levels encountered in fish range from 0.01–12 mg kg$^{-1}$. Methods are based on gas chromatography[70,74,118,152,145], high performance liquid chromatography[150,153], neutron activation analysis[103] (LD between 0.02 and 0.00004 mg kg$^{-1}$). A particularly sensitive method[152] is based on ethylation of alkylmercury with NaB(C$_2$H$_5$)$_4$ followed by cryogenic trapping and gas chromatography (LD 0.000004 mg kg$^{-1}$).

**Alkyl lead compounds**

Alkyl lead compounds have been encountered in crustacea and fish, respectively, in the concentration ranges 0.01–0.05 mg kg$^{-1}$ and 5–20 mg kg$^{-1}$. Atomic absorption spectrometric methods[201,202] with detection limits of 0.01–0.02 mg L$^{-1}$ are perfectly suitable for analysing alkyl lead compounds at the lowest levels encountered in these types of sample.

**Alkyltin compounds**

Alkyltin compounds have been encountered in invertibrates (oysters, mussels) in the range 0.02–0.1 mg kg$^{-1}$. Methods based on reduction of organotin to hydrides with sodium borohydride followed by atomic absorption spectrometry have detection limits as low as 0.000023 mg kg$^{-1}$.

**Organoarsenic compounds**

Organoarsenic levels encountered range from 0.00001–23 mg kg$^{-1}$ (molluscs) to 1.4–23 mg kg$^{-1}$ (prawn). The most sensitive method listed is high performance liquid chromatography with ICP mass spectrometric detector which has a detection limit of 0.002 mg kg$^{-1}$. This method is not sensitive enough to determine organoarsenic compounds at the 0.00001 mg kg$^{-1}$ level but will be adequate for many types of samples.

In the above discussion there are several instances quoted, mainly concerned with sediment analysis where it would appear that even the most sensitive published analytical methods would not be capable of determining some type of organometallic compounds at the lowest levels likely to be encountered in environmental samples, levels, which in some cases may be of toxicological significance. The quoted examples include alkylmercury compounds which can occur at concentrations down to 0.00001 mg kg$^{-1}$ in sediments the most sensitive published method being capable of determining only down to 0.0002 mg kg$^{-1}$. Whilst this method is probably adequate for the analysis of environmental materials, it may be that, at times, a fivefold more sensitive method is needed.

A further example concerns the determination of alkylead compounds, again in sediments. This can occur at 0.00001 mg kg$^{-1}$ in river and sea sediments, yet the most sensitive published method is capable of determining only down to 0.015 mg kg$^{-1}$.

Sensitive methods do not appear to be available for the determining of organoarsenic compounds in sediments and organoarsenic can be determined only in amounts down to 0.001 mg kg$^{-1}$ in biological samples such as fish or crustacea. Again, there may occasionally be a requirement to determine organoarsenic at levels down to 0.00001 mg kg$^{-1}$ in such samples.

As mentioned above, and discussed by the author[324,329] one solution to these problems is to adapt a sample preconcentration technique.

Thus, a procedure has been described by Lee[72] in which 20 L of water containing alkylmercury compounds is concentrated with sulfydryl cotton fibre. The alkylmercury compounds adsorb on to the fibre quantitatively. The cotton fibre is then removed and extracted with a few millilitres (say, 10 mL) of dilute hydrochloric acid/sodium chloride i.e. the preconcentration stage in which a concentration factor of 20.000/10 = 2000 is achieved. The alkyl mercury content of the extract is then determined by gas chromatography using an electron capture detector. A detection limit of 0.04 ng L$^{-1}$ alkylmercury is achieved in this procedure i.e. $0.04 \times 20/10^6 = 0.0000008$ mg absolute of alkylmercury can be determined.

If this procedure were applied to an aqueous extract of 100 g sediment then the detection limit for alkylmercury compounds would be 0.000008 mg 100 g$^{-1}$ i.e. 0.00008 mg kg$^{-1}$ which is fairly near to the lowest alkylmercury content likely to be encountered in sediment samples (0.00001 mg kg$^{-1}$).

A further example of the reduction of detection limits by the use of preconcentration is the determination of low levels of organoarsenic compounds. Persson and Irgum[321] preconcentrated sub ppm levels of dimethyl arsinite by passage of a 20 mL sample of seawater down a column of a strong cation exchange resin. (Dowex AG.50W-WB). The absorbed organoarsenic compound is then dissolved with a small volume (i.e. preconcentrated) and determined in the eluate by graphite furnace atomic absorption spectrometry. The detection limit of this method was 0.02 µg L$^{-1}$ organoarsenic, i.e. 0.0004 µg alkylarsenic absolute in a 20 mL seawater sample.

If this procedure were applied to a 20 mL extract of a 1 g sample of sediment then a detection limit of 0.0004 mg kg$^{-1}$ would be achieved (i.e. 0.0004 µg g$^{-1}$, 0.0000004 mg g$^{-1}$, 0.0004 mg kg$^{-1}$).

# REFERENCES

1. Kudo A., Nagasi H., Ose Y. *Water Research* **16** 1011 (1982)
2. Magos L. *Analyst (London)* **96** 847 (1971)
3. Kimura Y., Miller V.L. *Anal. Chim. Acta* **27** 325 (1962)

4. Environmental Protection Agency Publication No. EPA-625/6-74-003, p118, *Mercury in Fish Provisional Method*, Analytical Quality Control Laboratory Cinncinati, Ohio (1972)
5. Bennett T.B., McDaniel W.H., Hemphill R.N. *Advances in Automated Analysis* Technical International Congress, Vol 8, Mediad Inc, Tarrytown, N.Y. (1972)
6. El-Awady A.A., Miller R.B., Carter W.J. *Analytical Chemistry* **48** 110 (1976)
7. Umezaki Y., Iwamoto K. *Japan Analyst* **20** 173 (1971)
8. Rubel S., Lugowska M. *Anal. Chim. Acta* **115** 343 (1980)
9. Ke P.J., Thibert R.T. *Mikro Chimica Acta* **3** 417 (1973)
10. Chau Y.K., Saitoh H. *Environ. Sci. Technol.* **4** 839 (1970)
11. Braun T., Abbas M.M., Bakos L., Elek •. *Anal. Chim. Acta* **131** 311 (1981)
12. Muscat V.I., Vickers T.J., Andreay A. *Analytical Chemistry* **44** 218 (1972)
13. Olaffson J. *Anal. Chim. Acta* **68** 207 (1974)
14. Jones P., Nickless S.G. *Analyst (London)* **103** 1121 (1978)
15. Braun T., Abbas M.N., Torak S., Zvakefalvi-Nagy Z. *Anal. Chim. Acta* **160** 277 (1984)
16. Becknell D.E., March R., Allie W. *Analytical Chemistry* **43** 1230 (1971)
17. Robert J.M., Robenstein D.L. *Analytical Chemistry* **63** 2674 (1991)
18. Hanna C.P., Tyson J.F. *Analytical Chemistry* **65** 653 (1993)
19. Cullen M.C., McGuinness •. *Analytical Biochemistry* **42** 455 (1971)
20. Stary J., Prasilova J. *Radiochem. Radioanal. Lett.* **24** 143 (1976)
21. Stary J., Prasilova J. *Radiochem., Radioanal. Lett.* **26** 33 (1976)
22. Stary J., Prasilova J. *Radiochem., Radioanal. Lett.* **26** 193 (1976)
23. Stary J., Prasilova J. *Radiochem., Radioanal. Lett.* **27** 51 (1976)
24. Stary J., Havlik B., Prasilova J., Kratzer K., Hanusova J. *Int. J. Environ. Chem.* **5** 89 (1978)
25. Rains T.G., Menis O. *J. Assoc. Off. Anal. Chem.* **55** 1339 (1972)
26. Baltisberger R.J., Knaudson C.L. *Anal. Chim. Acta* **73** 265 (1974)
27. Stainton M.P. *Analytical Chemistry* **43** 625 (1971)
28. Nishi S., Horimoto H. *Japan Analyst* **17** 1247 (1968)
29. National Institute for Drug Abuse, Rockville Md, U.S.A. Research Monograph 21 (1978)
30. Dove A.K., Aronow R., Miceli J.M. *National Institute for Drug Abuse*, Rockville Md, USA **21** 210 (1978)
31. Doherty P.E., Dorsett R.S. *Anal. Abstr.* **43** 1887 (1971)
32. Simpson W.R., Nickless G. *Analyst (London)* **102** 86 (1977)
33. Lutze R.L. *Analyst (London)* **104** 979 (1979)
34. Goulden P.D., Anthony D.H.J. *Anal. Chim. Acta* **120** 129 (1980)
35. Kiemeneij, A.M., Kloosterboer, S.A. *Analytical Chemistry* **48** 575 (1976)
36. Farey B.J. Nelson, L.A. and Rolph, M.J. *Analyst (London)* **103** 656 (1978)
37. Minagawa K., Takizawa A., Kifune I. *Anal. Chim. Acta* **115** 103 (1980)
38. Yamagami E., Tateishi K., Hashimoto A. *Analyst (London)* **105** 491 (1980)
39. Oda C.E., Ingle J.D., *Analytical Chemistry* **53** 2305 (1981)
40. Oda C.E., Ingle J.D., *Analytical Chemistry* **53** 2030 (1981)
41. Hawley J.E., Ingle J.D.F. *Analytical Chemistry* **47** 719 (1975)
42. Christman D.R., Ingle J.D. *Anal. Chim. Acta* **86** 53 (1976)
43. Bisagni J.J., Lawrence A.W. *Environ. Sci. Technol.* **8** 850 (1974)
44. Carr R.A., Hoover J.B., Wilkniss P.W. *Deep Sea Res.* **19** 747 (1972)
45. Fitzgerald W.F., Lyons M.L. *Nature (London)* **242** 452 (1973)
46. Watling R.J., Watling H.M. *Water S.A.* **1** 113 (1975)
47. Abo-rady N.D.K. *Fresenius. Z. Anal. Chem.* **299** 187 (1979)
48. Jackson F., Dellar D. *Water Research* **13** 381 (1979)
49. Ahmed R., Stoeppler M. *Analyst (London)* **111** 1371 (1986)

50. Kimura O., Miller O. *Anal. Abstr.* **10** 2943 (1963)
51. Grantham D.L. *Laboratory Practice* **27** 294 (1978)
52. Graf E., Polos L., Bezur L., Pungor E. *Magy. Kem. Foly.* **79** 471 (1973)
53. Kalb G.W. *Atomic Absorption Newsletter* **9** 84 (1970)
54. Water Research Council and Department of the Environment (U.K.). *Determination of Organomercury Compounds in the Environment.* H.M. Stationary Office, London, pp 22 (1978)
55. Report Water Pollution Research Laboratory, Stevenage, U.K., Report No. 1272 (1972)
56. Goulden P.D., Afghan B.K. *Technical Bulletin No 27.* Inland Water Boards Department of Energy, Mines and Resources, Ottawa, Canada (1980)
57. Hatch W.R., Ott W.L. *Analytical Chemistry* **40** 2005 (1968)
58. Feldman C. *Analytical Chemistry* **46** 99 (1974)
59. World Health Organisation. *International standard for Drinking Water.* World Health Organisation, Geneva (1971)
60. World Health Organisation Official Journal of the European Community, *Proposal for a Council Directive Relating to the Quality of Water for Human Consumption.* **18** C214, 2-17 (1975)
61. Nishi S., Horimoto H. *Japan Analyst* **19** 1646 (1970)
62. Dressman R.C. *Journal of Chromatographic Science* **10** 472 (1972)
63. Ealy J.A., Shultz W.D., Dean J.A. *Anal. Chim. Acta* **64** 235 (1973)
64. Longbottom J.E. *Analytical Chemistry* **44** 1111 (1972)
65. Mushak P., Tibetts F.E., Zarneger P., Fisher G.B. *J. Chromatography* **87** 215 (1973)
66. Westhoo G. *Acta. Chem. Sci. Anal.* **21** 1790 (1967)
67. Westhoo G. *Acta. Chem. Sci. Anal.* **22** 2277 (1968)
68. Jones P., Nickless G. *J. Chromatography* **76** 285 (1973)
69. Jones p., Nickless G. *J. Chromatography* **89** 201 (1974)
70. Cappon C.J., Crispin Smith V. *J. Anal. Chem.* **49** 365 (1977)
71. Zarneger P., Mushak P. *Anal. Chim. Acta* **69** 389 (1974)
72. Lee Y.H. *International Journal of Environmental Analytical Chemistry* **29** 263 (1987)
73. Olson B.H., Cooper R.C. *Water Research* **10** 113 (1976)
74. Davies I.M., Graham W.C., Pirie J. *Marine Chemistry* **7** 111 (1979)
75. Compeau G., Bartha R. *Bulletin of Environmental Contamination and Toxicology* **31** 486 (1983)
76. Stoppler M., Matthas W. *Anal. Chim. Acta* **98** 389 (1978)
77. Sipos L., Nurenberg H.W., Valenta P., Branicia N. *Anal. Chim. Acta* **15** 25 (1980)
78. Olson K. *Analytical Chemistry* **49** 23 (1977)
79. Dean A., Rues R.E. *Analytical Department of Environment and Natural Water Board* UK H.M. Stationary Office, London. 22pp Ref. 17 (1978)
80. Millward G.E., Bihan A.T. *Water Research* **12** 979 (1978)
81. Agemian H., Chau V. *Anal. Chim. Acta* **101** 193 (1978)
82. Filippelli M. *Analyst (London)* **109** 515 (1984)
83. Tanabe K., Chiba K., Haraguchi H., Fuwa K. *Analytical Chemistry* **53** 1450 (1981)
84. Chiba K., Wong P.T.S., Tanabe K., Ozaki M., Haraguchi H., Winefordner J.D., Fuwa K. *Analytical Chemistry* **54** 761 (1982)
85. Yamamoto J., Kanada Y., Hisaka Y. *International Journal of Environmental Analytical Chemistry* **16** 1 (1983)
86. Department of the Environment at National Water Council, U.K. *Mercury in Waters, Effluents, Soils and Sediments. Additional Methods.* (PE-22-AGENW) H.M.S.O. (London) (1985)
87. Ahmed R., May K., Stoeppler M., *Fresenius Z. Anal. Chem.* **326** 510 (1987)
88. Murakiami T., Yoshinaga T. *Japan Analyst* **20** 1145 (1971)

89. Itsuki L., Komoro H. *Japan Analyst* **19** 1214 (1970)
90. Murakami T., Yoshinaga T. *Japan Analyst* **2** 878 (1971)
91. Carpenter W.L. *NCASI Stream Improvement Tech. Bull. No.* 263 (1972)
92. Takeshita R., Akagi H., Fujita M., Sakegami •. *J. Chromat.* **51** 283 (1970)
93. Omang S.H. *Anal. Chim. Acta* **52** 415 (1972)
94. Uthe J.E., Armstrong F.A.J., Stainton M.P. *J. Fish. Res. Board, Can.* **27** 805 (1970)
95. Osland R. *Pye Unicam Spectrovision No.* 24, 11 (1970)
96. Van Ettekovan K.G. $H_2O$ **13** 326 (1980)
97. Society for Analytical Chemistry. *Report of the Metallic Impurities Sub Committee of the Society for Analytical Chemistry* (1965)
98. Kimura Y, Miller V.L. *J. Agricultural Food Chemistry* **12** 253 (1964)
99. Polley D., Miller V.L. *Analytical Chemistry* **27** 1162 (1955)
100. Leong P.C., Ong H.O. *Analytical Chemistry* **43** 940 (1971)
101. Bretthauer E.W., Moghissi A.A., Snyder S.S., Matthews N.W. *Analytical Chemistry* **46** 445 (1974)
102. Anderson D.H., Evans J.H., Murphy J.J., White W.W. *Analytical Chemistry* **45** 1511 (1971)
103. Pillay K.K.S., Thomas C.C., Sonde C.J.A., Hyone C.M. *Analytical Chemistry* **43** 1419 (1971)
104. Bishop J.N., Taylor L.A., Neary B.O. *The Determination of Mercury in Environment.* US Environmental Protection Agency, Cincinnati, Ohio, P-120 (1975)
105. Environmental Protection Agency, *Methods for Chemical Analysis of Water and Wastes.* U.S. Environmental Protection Agency, Cincinnati, Ohio pp 134-138 (1974)
106. Iskandor I.K., Sayers J.K., Jakobs L.W., Keeney D.R., Gilmour J.T. *Analyst (London)* **97** 388 (1972)
107. Langmyhr F.J., Aamodt J. *Anal. Chim. Acta* **87** 483 (1976)
108. Matsunaga K., Takahasi *Anal. Chim. Acta* **87** 487 (1976)
109. Craig P.J., Morton S.F. *Nature (London)* **261** 125 (1976)
110. Official Methods of Analysis of AOAC (11th Edition) 418 (1976)
111. AOAC Offical Methods of the Association of Official Analytical Chemists, 11th Edition $_p$418 (1970)
112. Jirka A.M., Carter M.J. *Analytical Chemistry* **50** 91 (1978)
113. Matsumaya K., Takahasi S. *Anal. Chim. Acta* **87** 487 (1976)
114. Bartlett P.D., Craig P.J., Morton S.F. *Nature (London)* **267** 606 (1977)
115. Wanatabe N., Yasuda Y., Kato K., Nakamura T., Funasaka R., Shimokawa K., Sato E., Ose Y. *Science of the Total Environment* **34** 169 (1984)
116. Uthe J.F., Armstrong F.A.J., Tam K.C. *J. Ass. Official Analytical Chemists* **54** 866 (1972)
117. Longbottom J.E., Dressman R.C., Lichtenberg J.J. *J. Ass. of Official Analytical Chemists* **56** 1297 (1973)
118. Decadt G. Baeyens W.B., Bradley D., Goeyens L. *Analytical Chemistry* **57** 2785 (1985)
119. Guterman N.H., Lisk D.J. *Agricultural and Food Chemistry* **8** 306 (1960)
120. Houpt D.M., Campaan H. *Analysis* **1** 27 (1972)
121. Miller V.L., Lillis D. *Analytical Chemistry* **30** 1705 (1958)
122. Gage J.C. *Analyst (London)* **86** 457 (1961)
123. Ashley M.G. *Analyst (London)* **84** 692 (1959)
124. Collett D.L., Fleming J.E., Taylor G.E. *Analyst (London)* **105** 897 (1980)
125. Hendzel M.R., Jamieson D.E. *Analytical Chemistry* **48** 926 (1976)
126. Aspila K.A., Carron J.M. *Interlaboratory Quality Control Study* No. 18. Total Mercury in Sediments. Report Series, Inland Waters Directorate Water Quality Branch, Special Services Section, Department of Fisheries and Environment, Burlington, Ontario, Canada (1980)

127. Society for Analytical Chemistry (London). Report prepared by a sub committee of the Analytical Methods Committee and the Association of Official Analytical Chemists. The Determination of Mercury and Methylmercury in Fish. *Analyst (London)* **102** 769 (1977)
128. Davies I.M. *Anal. Chim. Acta* **102** 189 (1978)
129. Capelli R., Fezia C., Franchi A., Zaniochi O. *Analyst (London)* **104** 1197 (1979)
130. Shum G.T.C., Freeman H.C., Utha J.F. *Analytical Chemistry* **51** 414 (1979)
131. Schulz C.D., Clear D., Pearson J.E., Rivers J.B., Hylin J.W. *Bulletin of Environmental Contamination and Toxicology* **15** 230 (1976)
132. Agemian H., Chau A.S.Y. *Anal. Chim. Acta* **75** 297 (1975)
133. Holden A.V. *Pesticide Science* **4** 399 (1973)
134. Society for Analytical Chemistry (London) Report prepared by the metallic impurities in organic matter sub committee—The Use of Fifty Percent Hydrogen Peroxide for the Destruction of Organic Matter (Second Report). *Analyst (London)* **101** 62 (1976)
135. Hanamura S., Smith B.W., Winefondner J.D. *Analytical Chemistry* **55** 2026 (1983)
136. Beauchemin D., Sin K.W.M., Berman S.S. *Analytical Chemistry* **60** 2587 (1988)
137. Westhoo G. *Acta Chim. Sci. Anal.* **20** 2131 (1966)
138. Westhoo G., Johanssen B., Ryhage R. *Acta Chem. Scand.* **24** 2349 (1970)
139. Sjosfrand B. *Analytical Chemistry* **36** 814 (1964)
140. Bache C.A., Lisk D.J. *Analytical Chemistry* **43** 950 (1971)
141. Bye R., Paus P.E. *Anal. Chim. Acta* **107** 169 (1979)
142. Callum G.I., Ferguson M.M. Lenihan J.M.A. *Analyst (London)* **106** 1009 (1981)
143. Van Burg R., Farris F., Smith J.C. *J. Chromatography* **97** 65 (1974)
144. Odanaka Y., Tsuchlya W., Matono O., Goto S. *Analytical Chemistry* **55** 929 (1983)
145. Kamps L.R., McMahon I. *J. Ass. Official Analytical Chemists* **18** 351 (1970)
146. Newsom W.H., *J. Agricultural Food Chemistry* **19** 567 (1971)
147. Uthe J.F., Solomon J., Grift B. *J. Ass. Official Analytical Chemists* **55** 583 (1972)
148. Hight S.C. *J. Ass. Official Analytical Chemists* **70** 24 (1987)
149. Hight S.C. *J. Ass. Official Analytical Chemists* **70** 667 (1987)
150. Holak W., *Analyst (London)* **107** 1457 (1982)
151. Panoro K.W., Erikson D., Krull I.A. *Analyst (London)* **112** 1097 (1987)
152. Fisher R., Rapsomankis S., Andreae M.O *Analytical Chemistry* **65** 763 (1993)
153. MacCrehan W.A., Durst R.A. *Analytical Chemistry* **50** 2108 (1978)
154. Kanda Y., Suzuki N. *Analytical Chemistry* **52** 1672 (1980)
155. Miller V.L., Wachter L.E. *Analytical Chemistry* **22** 1312 (1950)
156. Stuart D.C. *Anal. Chim. Acta* **96** 83 (1978)
157. Christie A.A., Dundson A.J., Marshall B.S. *Analyst (London)* **92** 185 (1967)
158. Miller V.L., Polly D. *Analytical Chemistry* **26** 1333 (1954)
159. Kimura Y., Miller V.L. *Analytical Chemistry* **32** 420 (1960)
160. Miller V.L., Swanberg F. *Analytical Chemistry* **29** 391 (1957)
161. Long S.J., Scott D.R., Thompson R.I. *Analytical Chemistry* **44** 1111 (1973)
162. Dumareov R., Heindryck R., Dams K., Hoste J. *Anal. Chim. Acta* **107** 159 (1979)
163. Schroeder W.H., Jackson R. *Proceedings of the APCA Speciality Conference on Measurement and Monitoring of Non Criteria Contaminants in Air*. Chicago, March 22nd 1983. APCA Speciality Conference Proceedings SP-50, pp 91–100 (1983)
164. Aue W.A., Tell P.M. *J. Chromatography* **62** 15 (1971)
165. Torsi G., Desimoni E., Palmisano F., Sabbatini L. *Analytical Chemistry* **53** 1035 (1981)
166. Torsi G., Desimoni E., Palmisano F., Sabbatini L. *Analyst (London)* **107** 96 (1982)
167. Ballantine D.S., Zoller W.H. *Analytical Chemistry* **56** 1288 (1984)
168. Goulden P.D., Afghan B.K. *Technicon International Congress Vol II*, Nov 2–4, Futura, New York, p317 (1970)

169. Rosain R.M., Wai C.M. *Anal. Chim. Acta* **65** 279 (1973)
170. Carr R.A., Wilkniss P.E. *Environ. Sci. Technol.* **7** 63 (1973)
171. Gaston G.N., Lee A.K. *J. Am. Water Works Ass.* **66** 495 (1974)
172. Kopp J.F., Longbottom M.C., Lobring L.B. *J. Am. Water Works Ass.* **20** 64 (1972)
173. Coyne R.V., Collins J.A. *Analytical Chemistry* **44** 1093 (1972)
174. Bothner M.H., Robertson D.E. *Analytical Chemistry* **47** 592 (1975)
175. Weiss H.V., Chew K. *Anal. Chim. Acta* **67** 444 (1973)
176. Masri M.S., Friedman, *Environ. Sci. Technol.* **7** 951 (1973)
177. Newton D.W., Ellis J.R. *J. Environ. Qual.* **3** 20 (1974)
178. Jonasson I.R., Lynch J.J., Trip L.J. *Geol. Surv. Canada Paper* 73–21 (1973)
179. Toribara T., Shilds C.P., Koval L. *Talanta* **17** 1025 (1970)
180. Shimomura S., Kise A. *Bunseki Kagaku* **16** 1412 (1969)
181. Avotins P., Jenne E.A. *J. Environ. Qual.* **4** 515 (1975)
182. Lo J.M., Wai C.M. *Analytical Chemistry* **4** 1869 (1975)
183. Coe M., Cruz R., Van Loon J.C. *Anal. Chim. Acta* **120** 171 (1980)
184. Walton A.P., Wei, Guor Izo, Liang Z., Michel R.G., Morris I.B. *Analytical Chemistry* **63** 222 (1991)
185. Heiden R.W., Aikens D.A. *Analytical Chemistry* **49** 668 (1977)
186. McFarland R.C. *Radiochem. Radioanal. Lett.* **16** 47 (1973)
187. Coyne R.V., Collins J.A. *Analytical Chemistry* **44** 1093 (1972)
188. Carron J., Agemian H. *Anal. Chim. Acta* **92** 61 (1977)
189. Jin K., Shibata, Morita M. *Analytical Chemistry* **63** 986 (1991)
190. Giang G.B., Maxwell B., Siu K.W.M. *Analytical Chemistry* **63** 1506 (1991)
191. Glockling F. *Analytical Proceedings (London)* **17** 417 (1980)
192. Rapsomanikis S., Donard O.F.X., Weber J.H. *Analytical Chemistry* **58** 35 (1986)
193. Estes S.A., Uden P.C., Barnes R.M. *Analytical Chemistry* **53** 1336 (1981)
194. Forsythe D.D., Marshall W.D. *Analytical Chemistry* **55** 2132 (1983)
195. Hambrick G.A., Froelich P.N., Meinrate O.A., Lewis B.L. *Analytical Chemistry* **56** 421 (1984)
196. Braman R.S., Tompkins M.A. *Analytical Chemistry* **50** 1088 (1978)
197. Chau Y.K., Wong P.T.S., Goulden P.D. *Anal. Chim. Acta* **85** 421 (1976)
198. Chau Y.K., Wong P.T.S., Kramer O. *Anal. Chim. Acta* **146** 211 (1983)
199. De Jonghe W.R.A., Van Mol W.E., Adams F.C. *Analytical Chemistry* **55** 1050 (1983)
200. Chakraborti D., De Jonghe W.R.A., Van Mol W.E., Van Cleuvenbergen R.J.A., Adams F.C. *Analytical Chemistry* **56** 2692 (1984)
201. Chau Y.K., Wong P.T.S., Bengert G.A., Kramer O. *Analytical Chemistry* **51** 186 (1979)
202. Chau Y.K., Wong P.T.S., Saitoh H. *J. Chromatographic Science* **162** 14 (1976)
203. Segar D.A. *Analytical Lett.* **7** 89 (1974)
204. Estes S.A., Uden P.C., Barnes R.M. *Analytical Chemistry* **54** 2402 (1982)
205. Chau Y.K., Wong P.T.S., Bengert G.A., Dunn J.L. *Analytical Chemistry* **56** 271 (1984)
206. Hodges D.J., Naden F.C. *Presented at the International Conference on Heavy Metals in the Environment.* Society for Analytical Chemistry, London, pp 408–411 (1979)
207. Colombini M.P., Fuoco R., Papoff M. *Science of the Total Environment* **37** 61 (1984)
208. Potter M.R., Jarview A.W.P., Markell R.N. *Water Pollution Control* **76** 123 (1977)
209. Jarvis A.W.P., Whitmore A.P., Markall R.N., Potter H.R. *Environmental Pollution Series B* **6** 69 (1983)
210. Bond M.A., Bradbury J.R., Hanno P.J., Havell G.N., Hudson H.A., Strother S. *Analytical Chemistry* **56** 2392 (1984)
211. Lobinski R., Bontrom C.F., Candelone J.P., Hong S., Lobinska J.S., Adams F.C. *Analytical Chemistry* **65** 2510 (1993)

212. Blaszkewicz M., Baumhoer G., Neidbert B. *International Journal of Environmental Analytical Chemistry* **28** 207 (1987)
213. Imura S., Kukutako K., Aoki H., Sakai T. *Japan Analyst* **20** 704 (1971)
214. Aneva Z. *Fresenius Z. Anal. Chem.* **321** 680 (1985)
215. Reisinger K., Stoeppler M., Nurnberg H.N. *Nature (London)* **291** 228 (1981)
216. Sirota C.R., Uthe J.F. *Analytical Chemistry* **49** 823 (1977)
217. Harrison R.M., Laxen D.P.H. *Nature (London)* **275** 738 (1978)
218. Birnie S.E., Hodges D.J. *Environmental Technology Lett.* **2** 433 (1981)
219. Moss R., Browett E.V. *Analyst (London)* **91** 428 (1966)
220. Hancock S., Slater A. *Analyst (London)* **100** 422 (1975)
221. De Jonghe W., Adams F.C. *Anal. Chim. Acta* **108** 21 (1979)
222. Thilliez G. *Analytical Chemistry* **39** 427 (1967)
223. Torsi G., Palmisamo F. *Analyst (London)* **108** 1318 (1983)
224. Soulages N.L. *Analytical Chemistry* **38** 28 (1966)
225. Reamer D.O., Zoller W.H., O'Haver T.C. *Analytical Chemistry* **50** 1449 (1978)
226. De Jonghe W., Chakraborti D., Adams F. *Anal. Chim. Acta* **115** 89 (1980)
227. De Jonghe W., Chakrabarti D., Adams F.C. *Analytical Chemistry* **52** 1974 (1980)
228. Boettner E.A., Dallas F.C. *J. Gas Chromatography* **190** (1965)
229. Radzuik B., Thomassen V., Van Loon J.C., Chau Y.K. *Anal Chim. Acta* **105** 255 (1979)
230. Robinson T.W., Kiesch E.L., Goodbread J.P., Bliss R., Marshall R. *Anal. Chim. Acta* **92** 321 (1977)
231. Koizumi H., McLaughlin R.D., Hadeishi T. *Analytical Chemistry* **51** 387 (1979)
232. Luskina B.M., Syavtsillo S.V. *Nov. Obl. Prom. Sauit. Khim.* **186** (1969)
233. Coyle C.F., White C.E. *Analytical Chemistry* **29** 1486 (1957)
234. Vernon F. *Analytical Chemistry* **46** 29 (1974)
235. Blunden S.J., Chapman A.H. *Analyst (London)* **103** 1266 (1978)
236. Aldridge W.N., Cremer J.E. *Analyst (London)* **82** 37 (1957)
237. Chau Y.K., Wong P.T.S., Bengert G.A. *Analytical Chemistry* **54** 246 (1982)
238. Hodge V.F., Seidel S.L., Goldberg E.D. *Analytical Chemistry* **51** 1256 (1979)
239. Meinema H.A., Burger W.T., Verslius-Dehaan G., Gevers E.C. *Environmental Science and Technology* **12** 288 (1978)
240. Unger M.A., MacIntyre W.G., Greaves J., Huggett R.J. *Chemosphere* **15** 461 (1986)
241. Soderguist C.J.. Crosby D.C. *Analytical Chemistry* **50** 1435 (1978)
242. Hattori Y., Kobayashi A., Takemoto S., Takami K., Kuge Y., Sigimae A., Naramoto N. *J. Chromatography* **315** 341 (1984)
243. Neubert G., Wirth H.O. *Z. Anal. Chem.* **273** 19 (1975)
244. Neubert G., Andreas H. *Z. Anal. Chem.* **280** 31 (1976)
245. Braman R.S., Tompkins M.A. *Analytical Chemistry* **51** 12 (1979)
246. Jackson J.A., Brinckmann F.E., Iverson W.P. *Environmental Science and Technology* **16** 110 (1982)
247. Huey C., Brinckmann F.E., Iverson W.P. In *Proceedings of the International Conference on the Transport of Persistant Chemicals in Aquatic Ecosystems* (Editors A.S.W., Freitas, D.J. Kushner, D.S.U. Quadri), National Research Council of Canada, Ottawa, Canada. pp 11.73–11.78 (1974)
248. Aue W.A., Flinn G.C. *J. Chromatography* **142** 145 (1977)
249. Nelson J.D., Blair W., Brinckmann F.E. *Applied Microbiology* **26** 321 (1973)
250. Lobinski R., Dirk W.M.R., Ceulemans M., Adams F.C. *Analytical Chemistry* **64** 159 (1992)
251. Matthias C.L., Bellamy J.M., Olson G.J., Brinckmann F.E. *Environmental Science and Technology* **20** 609 (1986)
252. Maguire R.J., Hunealt H. *J. Chromatography* **209** 458 (1981)
253. Mueller M.E. *Fresenius Z. Anal. Chem.* **317** 32 (1984)

254. Muller M.D. *Analytical Chemistry* **59** 617 (1987)
255. Colby S.M., Stewart M., Reilly J.P. *Analytical Chemistry* **62** 2400 (1990)
256. Nygren O., Nilsson C.A., Frech W. *Analytical Chemistry* **60** 2204 (1988)
257. Shen. L., Vela N.P., Sheppard B.S., Carns J.S. *Analytical Chemistry* **63** 1491 (1991)
258. Krull I.S., Panaro K.W. *Applied Spectroscopy* **39** 960 (1985)
259. Andreae M.O., Asmode J.F., Foster P., Van't dack L. *Analytical Chemistry* **53** 1766 (1981)
260. Waggon H., Jehle D. *Die Nahrung* **17** 739 (1973)
261. Waggon H., Jehle D. *Die Nahrung* **19** 271 (1975)
262. Odenaka Y., Matano O, goto S. *J. Agricultural and Food Chemistry* **26** 505 (1978)
263. Valkirs A.D., Seligman P.F., Olsen G.J., Brinckman F.E. *Marine Pollution Bulletin* **17** 320 (1986)
264. Valkirs A.O. Seligman P.F., Stang P.M. *Marine Pollution Bulletin* **17** 319 (1986)
265. Brinckmann F.E. In *Trace Metals in Seawater. Proceedings of a NATO Advanced Research Institute on Trace Metals in Seawater* 30/3-3/4/81, Sicily, Italy (Editors C.S. Wong et al.) Penum Press, New York (1981)
266. Ebdon L., Alonso J.I.G. *Analyst (London)* **112** 1551 (1987)
267. Laughlin R.B., Guard H.E., Coleman W.M. *Environmental Science and Technology* **20** 201 (1986)
268. Epler K.S., O'Haver T.C., Turk G.C., MacCrehan W.A. *Analytical Chemistry* **60** 2062 (1988)
269. Stephenson M.D., Smith D.R. *Analytical Chemistry* **60** 696 (1988)
270. Cal Y., Aizago R., Bayona J.M. *Analytical Chemistry* **66** 1161 (1994)
271. Gilmour C.C., Tuttle J.H., Means J.C. *Analytical Chemistry* **58** 1848 (1986)
272. Randall L., Han J.S., Weber J.H. *Environmental Technology Lett.* **7** 571 (1986)
273. Sinex S.A., Cantillo A.Y., Helz G.R. *Analytical Chemistry* **52** 2342 (1980)
274. Arakawa Y., Wada O., Manabe M. *Analytical Chemistry* **55** 1901 (1983)
275. Han J.S., Weber J.H. *Analytical Chemistry* **60** 316 (1988)
276. Smith J.D. *Nature (London)* **225** 103 (1970)
277. Short J.W. *Bulletin of Environmental Contamination and Toxicology* **39** 412 (1987)
278. Sasaki K., Ishiaku T., Susuki T., Saito Y. *J. Ass. of Official Analytical Chemists* **71** 360 (1988)
279. Arakawa Y., Wade Yu T.H., Iwai H. *J. Chromatography* **216** 209 (1981)
280. Sullivan J.J., Torrkelson J.D., Nickell H.M., Hollingworth T.A., Saxton W.L., Miller G.A. *Analytical Chemistry* **60** 626 (1988)
281. Gauer W.O., Seiber J.N., Crosby D.C. *J. Agricultural and Food Chemistry* **22** 252 (1974)
282. Francois R., Weber J.H. *Marine Chemistry* **5** 279 (1988)
283. Nangniot P., Martens P.H. *Anal. Chim. Acta* **24** 276 (1961)
284. Kopp J.F. *Analytical Chemistry* **45** 1786 (1973)
285. Kolthoff I.M., Belcher R. In *Volumetric Analysis* Volume 3 pp 511–513 Interscience Publishers, New York (1957)
286. Armstrong F.AJ., William P.L., Strickland J.D. *Nature (London)* **211** 481 (1966)
287. Manning D.C. *Atomic Absorption Newsletter* **10** 6 (1971)
288. Stringer C.E., Attrep M. *Analytical Chemistry* **51** 731 (1979)
289. Peoples S.A., Lakso J., Lais T. *Proc. Wst. Pharmacol. Soc.* **14** 178 (1971)
290. Haywood M.G., Riley J.P. *Anal. Chim. Acta* **85** 219 (1975)
291. Kamada T. *Talanta* **23** 835 (1976)
292. Sandhu S.S. *Analyst (London)* **101** 856 (1976)
293. Ishman M., Spencer R. *Analytical Chemistry* **49** 1599 (1977)
294. Braman R.S., Foriback C.C. *Science* **182** 1247 (1973)
295. Edmonde J.S., Francesconi K.A. *Analytical Chemistry* **48** 2019 (1976)

296. Braman R.S., Johnson D.L., Craig C., Foreback C.C., Ammons J.M., Bricker J.L. *Analytical Chemistry* **49** 612 (1977)
297. Gifford P.R., Bruckenstein S. *Analytical Chemistry* **52** 1028 (1980)
298. Talmi Y., Bostick D.T. *Analytical Chemistry* **47** 2145 (1975)
299. Andreae M.O. *Analytical Chemistry* **48** 820 (1977)
300. Crecelius E.A. *Analytical Chemistry* **50** 826 (1978)
301. Shalk A.U., Tallman D.E. *Anal. Chim. Acta* **98** 251 (1978)
302. Von Endt D.W., Kearny P.C., Kaufman D.D. *J. Agric. Food Chem.* **16** 17 (1968)
303. Portman J.E., Riley J.P. *Anal. Chim. Acta* **31** 509 (1964)
304. Casvalho M.B., Hercules D.M. *Analytical Chemistry* **50** 2030 (1978)
305. Lodmell J.D. *PhD Theses*, University of Tennessee, Knoxville, Tenn. (1973)
306. Johnson L.D., Gerhart K.O., Aue W.A. *Sci. Total Environ.* **1** 108 (1972)
307. Fickett A.W., Doughtrey E.H., Mishak P. *Anal. Chim. Acta* **79** 93 (1975)
308. Soderquist C.J., Crosby D.G., Bowers J.B. *Analytical Chemistry* **46** 155 (1974)
309. Parris G.E., Blair W.R., Brinckmann F.E. *Analytical Chemistry* **49** 378 (1977)
310. Lussi Schlatter B., Braudenberger H. In *Advances in Mass Spectrometry in Biochemistry and Medicine* pp 231-243. Spectrum Publications, New York (1976)
311. Jyn Jynn Yu, Wai C.M. *Analytical Chemistry* **63** 842 (1991)
312. Dietz E.A., Perez M.E. *Analytical Chemistry* **48** 1088 (1976)
313. Johnson D.L., Pilson M.E.O. *Anal. Chim. Acta* **58** 289 (1972)
314. Henry F.T., Thorpe T.M. *Analytical Chemistry* **52** 80 (1980)
315. Blais J.S., Momplasir G.M., Marshall W.P. *Analytical Chemistry* **62** 1611 (1990)
316. Grabinski A.A. *Analytical Chemistry* **53** 966 (1981)
317. Aggett J., Kadwani R. *Analyst (London)* **108** 1495 (1983)
318. Elton R.K., Geiger Jr. W.E. *Analytical Chemistry* **50** 712 (1978)
319. Henry F.T., Kirch T.O., Thorpe T.M. *Analytical Chemistry* **51** 215 (1979)
320. Myers D.J., Osteryoung J. *Analytical Chemistry* **45** 267 (1973)
321. Persson J., Irgum K. *Anal. Chim. Acta* **138** 111 (1982)
322. Yamamoto M., *Soil Sci. Soc. Am. Proc.* **39** 859 (1975)
323. Penrose W.R. *Critical Reviews of Environmental Control* **4** 465 (1974)
324. Crompton T.R, In *The analysis of Natural Waters* Vol 1, Complex Formation, Oxford University Press (1993)
325. Braman R.S., Johnson D.L., Forback C.O. *Analytical Chemistry* **49** 621 (1977)
326. Haward A.G., Arbab-Zavor M.H. *Analyst (London)* **106** 213 (1981)
327. Hinners R.A. *Analyst (London)* **106** 213 (1980)
328. Pierce F.D., Brown H.R. *Analytical Chemistry* **49** 1417 (1977)
329. Crompton T.R. In *The analysis of natural waters*, Vol 2, Direct Preconcentration, Oxford University Press, (1993)
330. Henderson S.L., Suyder L.J., *Analytical Chemistry* **33** 1172 (1961)
331. Maher W.A. *Anal. Chim. Acta* **126** 157 (1981)
332. Sandhu S.S., Nelson P. *Analytical Chemistry* **50** 322 (1978)
333. Woolson E.A., Kearny P.C. *Environ. Sci. Technol.* **7** 47 (1973)
334. McBride B.C., Wolfe R.S. *Biochemistry* **10** 4312 (1971)
335. Challenger F. *Chem. Rev.* **36** 315 (1945)
336. Franscesconi K.A., Hicks P., Stockton R.A., Irgolic K.J. *Chemosphere* **14** 1443 (1985)
337. Schwedt E.G., Russel H.A. *Chromatographia* **5** 242 (1972)
338. Beauchemin D., Bednas M.E., Berman S.S., McLaren J.W., Siu K.W.M., Sturgeon R.E. *Analytical Chemistry* **60** 2209 (1988)
339. Stockton R.A., Irgolic K. *J. Environmental and Analytical Chemistry* **6** 313 (1979)
340. White J.N.C., Englar J.R. *Botanica Marina* **26** 159 (1983)
341. Desada T.A., Ibrahim L.F. *Analyst (London)* **112** 549 (1987)
342. Killa H.M.A., Robenstein D.L. *Analytical Chemistry* **60** 2283 (1988)

343. Tanzer D., Henmann K.G. *Analytical Chemistry* **63** 1984 (1991)
344. Blotcky A.J., Ebrahim A., Rack E.P. *Analytical Chemistry* **60** 2734 (1984)
345. Andreae M.O., Froelich P.N. *Analytical Chemistry* **53** 287 (1981)
346. Sohrin Y. *Analytical Chemistry* **63** 811 (1991)
347. Jones A. *Analytical Chemistry* **66** 271 (1994)
348. Aue W.A., Miller B., Xun, Yun Sen. *Analytical Chemistry* **62** 2453 (1990)
349. Braun L., Haswell S.J., Rhead M.M., O'Neill P., Brancroft C.C. *Analyst (London)* **108** 1511 (1983)
350. Campana J.E., Risby T.H. *Analytical Chemistry* **52** 468 (1980)
351. Mahone L.G., Garner B.J., Buch R.R., Lane T.H., Tatera J.F., Smith R.C., Frye C.L. *Environmental Toxicology and Chemistry* **2** 307 (1983)
352. Van der Post D.C. *Water Pollution Control* **77** 520 (1978)
353. Horner H.J., Weiler J.A., Angelotti N.C. *Analytical Chemistry* **32** 858 (1960)
354. Barth P.J., Adams D.M., Soloway A.H., Mechetner E.B., Alam F.A. *Analytical Chemistry* **63** 890 (1991)
355. Kolesar E.S., Walsar P.M. *Analytical Chemistry* **60** 1731 (1988)
356. Kolesar E.S., Walsar P.M. *Analytical Chemistry* **60** 1737 (1988)
357. Grate J.W., Ros Perrsson S.L., Venezky D.L., Klusty M., Wohitjen H. *Analytical Chemistry* **65** 1868 (1993)
358. Pongratz R., Henmann K.G. *Analytical Chemistry* **68** 1262 (1996)

# Index

Air
  lead in 31
  organoarsenic in 31, 32
  organolead in 31, 32, 158–165
  organomanganese in 31, 32, 207
  organomercury in 31, 32, 133–135
  organonickel in 31, 32, 208–209
  organotin in 31, 32
Algae
  arsenic in 26, 27
  copper in 26, 27
  manganese in 26, 27
Algae, bioaccumulation of
  arsenic 86, 87
  copper 86, 87
Annelids, toxicity data 36
Antimony, concentration factor,
  water/sediment 77
Antimony in,
  air 31
  coastal waters 7, 15
  crops 28, 29
  food 51–53
  land 28, 29
  rivers 7, 15
  seawater 15
  sediments 7, 15
  sewage 29, 30
  soil 30
Aquatic organisms, det of organomercury
  119, 120
Arsenic, bioaccumulation in
  algae 82–87
  sediment 82–87
Arsenic in
  algae 26, 27
  coastal waters 198
  crustacea 25
  fish 24
  land 28, 29
  rivers 5, 7, 15, 27
  seawater 15
  sediments 5, 7, 15
  sewage 30
  soil 30

Atomic absorption on spectroscopy,
  determination of
  organoantimony 204
  organoarsenic 189–192, 197, 198, 202, 203
  organolead 139, 140, 151–154, 159
  organomercury 93–103, 106–110, 116, 117, 119, 121–123
  organotin 181, 183, 185, 186

Barnacle, bioaccumulation of zinc 86, 87
Bioaccumulation in
  fish 85–87
  invertebrates 88
  kelp 86, 87
  plants 88
Bioaccumulation of
  arsenic 82
  copper 79–82
  mercury 82, 83
  organolead 83
  organomercury 83, 84
  organotin 84, 85
Biological materials
  organoarsenic in 201–203
  organolead in 153–158
  organomercury in 119–133
  organoselenium in 204, 205
  organotin in 184–186

Cadmium, bioaccumulation in clams 86, 87
Clam, bioaccumulation of cadmium 86, 87
Coastal waters
  antimony in 15
  arsenic in 15
  copper in 15
  germanium in 15
  lead in 15
  mercury in 6–10, 15
  nickel in 6–10, 15
  organoarsenic in 7, 14
  organolead in 6–10, 14, 41
  organomercury in 7, 14
  organotin in 6–10, 14

# INDEX

Coastal waters (*cont.*)
  selenium in 15
  tin in 15
Coastal waters, concentration factors of
  metals 9, 10
  organometallics 9, 10
Concentration factors, water/sediments of
  antimony 9, 10, 76, 77, 78
  arsenic 9, 10, 76, 77, 78
  copper 9, 10, 76, 77, 78
  lead 70, 86, 94
  mercury 9, 10, 76, 77, 78
  manganese 9, 10, 76, 77, 78
  nickel 9, 10, 76, 77, 78
  selenium 9, 10, 76, 77, 78
  tin 9, 10, 76, 77, 78
Concentration factors, water/sediment in
  organoarsenic 9, 10, 76, 77, 78
  organolead 9, 10, 76, 77, 78
  organomercury 9, 10, 76, 77, 78
  organotin 9, 10, 76, 77, 78
Copper, bioaccumulation 77–83
Copper, bioaccumulation in
  algae 86, 87
  barnacle 86, 87
  muscle 86, 87
  opercle 86, 87
  phytoplankton 86, 87
  sediments 76, 86, 87
Copper in
  air 31, 32
  algae 26, 27
  foods 49–53
  coastal waters 7, 15
  crops 28, 29
  crustacea 24, 25
  fish 24, 25
  land 28, 29
  phytoplankton 26, 27
  rivers 5, 7, 15, 27
  seawater 13, 15
  sediments 5, 7, 15
  sewage 28–30
  soil 28–30
Copper, toxicity data 36
Crops
  antimony in 28, 29
  arsenic in 28, 29
  copper in 28, 29
  lead in 28, 29
  manganese in 28, 29
  mercury in 28, 29
  nickel in 28, 29
  organoarsenic in 28, 29, 203, 204
  organomercury in 28, 29, 119, 120
  organotin in 28, 29, 186, 187
  selenium in 28, 29
  tin in 28, 29
Crustacea
  arsenic in 25, 26
  copper in 25, 26
  cumulative LC 68, 70
  $LC_{50}$ 62–64
  $LE_{50}$ 62–64
  lead in 25, 26
  mercury in 25, 26
  nickel in 28, 29
  organoarsenic in 22, 25, 26
  organolead in 22, 25, 26
  organomercury in 22, 25, 26
  organotin in 22, 25, 26
  selenium in 25, 26
  toxicity data 36
Cumulative $LC_{50}$, determination of 68, 70

Differential pulse anodic scanning voltammetry of
  organoarsenic 197, 199–201
  organocadmium 187
  organolead 158

Eihinoderm, toxicity data 36
Emission spectrometry, determination of
  organogermanium 205
  organomercury 123

Fish
  arsenic in 24, 26, 27
  copper in 24, 26, 27
  cumulative $LC_{50}$ 68, 70
  $LC_{50}$ 47, 82–84
  lead in 24
  manganese in 24
  max. safe concentration standard 64
  mercury in 24, 26, 27
  nickel in 24
  organoarsenic in 20–24
  organolead in 20–24
  organomercury in 20–24
  organotin in 20–24
  selenium in 24
  toxicity data 36
Fish, bioaccumulation of
  general 86, 87
  lead 86, 87
  mercury 86, 87
  metals 85, 88
  organolead 86, 87

INDEX

organometallics 86, 87
Foods, metals in 49–53
Fourier transform infrared spectroscopy of organonickel 208

Gastropods, toxicity data 36
Germanium in
  coastal waters 15
  rivers 5, 15
  seawater 15
  sediments 5, 7, 15
Gas chromatography of
  organoarsenic 192–196, 198, 199, 202, 203
  organolead 139–164
  organomanganese 206–207
  organomercury 102–105, 108, 109, 111, 117–120, 124–132, 134, 135
  organosilicon 209
  organotin 166–177, 181–186

High performance liquid chromatography of
  organoarsenic 196
  organocopper 207, 208
  organolead 150, 151, 163–165
  organomanganese 206
  organomercury 131
  organotin 174–176, 177, 181–183, 186
  organoselenium 204
Hydride generation atomic absorption spectrometry of
  organogermanium 205, 206
  organotin 183
Hydride generation, gas chromatography of organotin 177–181

Inductively coupled atomic emission spectrometry of
  organomercury 123, 124
  organosilicon 209
Industrial effluents
  organoarsenic in 199–201
  organolead in 151
  organomercury in 110, 111
Invertebrates, bioaccumulation of
  metals 87
  organometallics 87
Ion exchange chromatography of
  organoarsenic 196, 197, 203
Iron, bioaccumulation in kelp 86, 87

Kelp, bioaccumulation of metals 86, 87

Land
  arsenic in 28, 29
  antimony in 28, 29
  copper in 28, 29
  lead in 28, 29
  manganese in 28, 29
  mercury in 28, 29
  selenium in 28, 29
  tin in 28, 29
Lead in
  air 31
  coastal waters 15
  crops 28, 29
  crustacea 25, 26
  fish 22, 24
  food 51–53
  land 5, 7, 15
  rivers 5, 7, 15
  seawater 13–15
  sediments 5, 7, 15
  sewage 28–30
  soil 28, 29
Lead
  bioaccumulation 86, 87
  concentration factor, water/sediments 76
  toxicity data 36
$LC_{50}$
  effect of experimental parameters 64
  test details 61–64

Manganese concentration factors, water/sediments 77
Manganese in
  algae 26–28
  coastal waters 7, 15
  crops 28, 29
  fish 24
  food 51–53
  land 28, 29
  phytolankton 26, 27
  rivers 5, 7, 15
  seawater 13, 15
  sediments 5, 7, 15
  sewage 36
  soil 36
Maximum safe concetration standard, details of 64–68
Mercury, bioaccumulation of
  fish 86, 87
  general 79–85
  mussel 86, 87
  phytoplankton 86, 87

Mercury, bioaccumulation of (cont.)
  sediments 86
Mercury, concentration factor,
  water/sediments 86
Mercury in
  air 30, 31
  coastal waters 7, 15
  crops 28, 29
  crustacea 25
  fish 24
  food 49–53
  land 28, 29
  rivers 5, 7, 15, 27
  seawater 13, 15
  sediments 5, 7, 28, 29
  soil 29, 30
  sewage 29, 30
Mercury, toxicity data 36
Molluscs
  $LC_{50}$ 41, 47
  toxicity data 36
Mussels, bioaccumulation of
  copper 86, 87
  mercury 86, 87
  organomercury 86, 87

Natural waters
  organoantimony in 204
  organoarsenic in 187–197
  organocopper in 207, 208
  organogermanium in 205, 206
  organolead in 40–42, 139–150
  organomanganese in 206
  organomercury in 91–105
  organotin in 165–178
  organoselenium in 204
Neutron activation analysis of
  organomercury 92, 93
Nickel, concentration factor, water/sediment
  76
Nickel in
  air 31, 32
  coastal waters 7, 15
  crops 28, 29
  fish 24
  food 51–53
  land 28, 29
  rivers 5, 7, 15
  seawater 13, 15
  sediments 5, 7, 15
  sewage 28–30
  soil 28, 29
Nickel, toxicity data 36
North Sea, pollution load 12, 16–20

Nuclear magnetic resonance spectroscopy
  of
  organomercury 92, 93, 119
  organoselenium 205
  organotin 182

Opercle, bioaccumulation of copper 86, 87
Organoantimony, determination in
  natural waters 204
Organoarsenic
  bioaccumulation in sediments 86, 87
  concentration factor, water/sediment 76
  toxicity of 48, 49
Organoarsenic in
  air 31, 32
  biological materials 202, 203
  coastal waters 6, 7, 8, 15
  crops 28, 29, 203, 204
  crustacea 22, 25, 26
  fish 24
  industrial effluents 199–201
  natural waters 187–197
  phytoplankton 26–28
  plants 203, 204
  potable water 30, 31, 187–197
  rivers 14
  seawater 11, 12, 14, 18, 198
  sediments 7, 14, 198, 199
  soil 28-30, 198, 199
  waste waters 199
Organoboron, determination in tissue 210
Organocadmium, determination in seawater
  187
Organocopper, determination in natural
  waters 207, 208
Organogermanium, determination in natural
  waters 205, 206
Organolead
  bioaccumulation in fish 83, 86, 87
  bioaccumulation in sedimenst 86, 87
  concentration factors, water/sediment 76
  toxicity of 40–42
Organolead in
  air 158–165
  biological materials 153–158
  coastal waters 6–10, 12–14
  crops 28, 29
  crustacea 22, 25, 26
  fish 20, 22, 24, 26–28
  industrial effluents 151–153
  natural waters 139–150
  phytoplankton 26–28
  plants 158
  potable water 139–150

rain 150, 151
rivers 5–9, 14
seawater 10, 12, 14, 150
sediments 5, 7, 14, 41, 152, 153
snow 150, 151
soil 28
waste waters 151
Organomanganese in
 air 207
 natural waters 206
Organomercury
 bioaccumulation of 83, 84
 in mussels 86, 87
 in sediments 86, 87
 concentration factor, water/sediment 76
 preservation of samples 135–139
 toxicity 35–40
Organomercury in
 air 138–135
 aquatic organisms 119, 120
 biological materials 120–133
 coastal waters 6–10, 14
 crops 118, 120
 crustacea 22, 25–27
 fish 22–26
 industrial effluents 110, 111
 natural waters 91–105
 phytoplankton 26–28
 plants 119, 120
 potable water 30, 31, 91–105
 rain 110
 rivers 5, 6, 9, 14, 91–105
 seawater 10–12, 14, 23, 24, 105–110
 sediments 5, 7, 14, 113–119
 sewage effluents 112, 113
 soil 27–29, 113–119
 waste waters 110, 111
Organonickel, determination in air 208, 209
Organophosphorus, determination of 211
Organoselenium in
 biological materials 204
 natural waters 205
Organosulphur, determination of 211
Organotin
 bioaccumulation 84, 85
 concentration factor, water/sediment 76
 $LC_{50}$ 45–47
 toxicity 42–47
Organotin in
 air 31, 32
 biological materials 184–186
 coastal waters 6–10, 12, 14
 crops 186, 187
 crustacea 22, 25, 26

fish 20, 22, 25, 26
natural waters 164–178
plants 186, 187
potable water 164–178
rain 182–184
rivers 5–9, 14
seawater 6, 10–16, 20, 21, 178–182
sediments 5, 7, 182–184
sewage 184
soil 28


Phytoplankton
 arsenic in 26, 27
 copper in 26, 27
 manganese in 26, 27
 mercury in 26, 27
 organoarsenic in 26–28
 organolead in 26–28
 organomercury in 26–28
 organotin in 26, 28
Phytoplankton, bioaccumulation of
 copper 86, 87
 mercury 86, 87
Plants
 organoarsenic in 203, 204
 organolead in 158
 organomercury in 118, 119
 organotin in 186, 187
Plants, bioaccumulation of
 metals 88
 organometallics 88
Polarography of
 organoantimony 204
 organolead 149
Potable water
 organoarsenic in 30, 31, 187–198
 organolead in 30–31, 139–150
 organomercury in 30, 31, 91–105
 organotin in 30, 31, 165–178
Preservation of organomercury 135–139

Rain
 organolead in 150, 151
 organomercury in 110, 111
 organotin in 182
Rivers
 antimony in 5, 7, 15
 arsenic in 5, 7, 15, 26, 27
 copper in 5, 7, 15
 lead in 5, 7, 15
 manganese in 5, 7, 15
 mercury in 5, 7, 15, 26, 27
 nickel in 5, 7, 15

Rivers (cont.)
  organoarsenic in 5, 6, 7, 14
  organolead in 5–8, 14
  organomercury in 5–8, 14
  organotin in 5–8, 14
  selenium in 5, 7
  tin in 5, 7, 15
Rivers, concentration factors,
    water/sediments of
  metals 9, 10
  organometallics 9, 10

Seawater
  antimony in 15
  arsenic in 15
  copper in 13, 15
  germanium in 15
  lead in 13, 15
  manganese in 13, 15
  mercury in 13, 15
  nickel in 13, 15
  organoarsenic in 10–12, 14, 17–19,
    198
  organocadmium 187
  organolead in 10–12, 17–20, 22, 150
  organomercury in 10–12, 14, 17, 22,
    105–110
  organotin in 8–12, 14, 20, 170–182
  selenium in 13, 15
  tin in 13, 15
Sediments
  antimony in 5, 7, 15
  arsenic in 5, 7, 15
  copper in 5, 7, 15
  germanium in 5, 7, 15
  lead in 5, 7
  manganese in 5, 7, 15
  mercury in 5, 7, 15
  nickel in 5, 7, 15
  organoarsenic in 14
  organolead in 5, 7, 14, 41, 152, 153
  organomercury in 5, 7, 14
  organotin in 5, 7, 14, 182–184
  selenium in 5, 7, 15
  tin in 5, 7, 15
Sediments, bioaccumulation of
  arsenic 86, 87
  copper 86, 87
  lead in 86, 87
  mercury 86, 87
  organoarsenic 86, 87
  organolead 86, 87
  organomercury 86, 87
Selenium in
  air 31, 32
  coastal waters 15
  crops 28, 29
  crustacea 25
  fish 24
  foods 51–53
  land 28, 29
  rivers 5, 7, 15
  seawater 13, 15
  sediments 5, 7, 15
  sewage 28–30
  soil 28–30
Selenium, concentration factor,
    water/sediments 87
Sewage
  antimony in 29, 30
  arsenic in 29, 30
  copper in 29, 30
  lead in 29, 30
  manganese in 29, 30
  mercury in 29, 30
  organomercury in 112, 113
  organotin in 184–186
  nickel in 29, 30
  selenium in 29, 30
  tin in 29, 30
Snow, determination of organolead 150,
    151
Soil
  antimony in 29, 30
  arsenic in 29, 30
  copper in 29, 30
  lead in 29, 30
  manganese in 29, 30
  mercury in 29, 30
  nickel 29, 30
  organoarsenic in 198, 199
  organomercury in 113–119
  selenium in 29, 30
  tin in 29, 30
Spectrofluorimetry of organotin 165, 166,
    178, 185
Spectrophotometric determination of
  organoarsenic 188–192, 199, 203
  organolead 151
  organomercury 92, 111–115, 119, 120
  organotin 165
Substoichiomtric analysis of organomercury
    132
Supercritical fluid chromatography of
    organotin 177, 183

Thin layer chromatography of
  organolead 149–150

INDEX

organomercury 110–111, 132
organotin 177, 178
Tin in
  air 31, 32
  coastal waters 15
  crops 28, 29
  land 28, 29
  rivers 5, 7, 15
  seawater 13, 15
  sediments 5, 7, 15
  sewage 28, 29
  soil 29, 30
Tin, concentration factor, water/sediment 76
Tin compounds, methylation of 42–44
Tissue, determination of organoboron 210
Toxicity data
  copper 36
  lead 36
  mercury 36
  nickel 36
  organoarsenic 48, 49
  organolead 40–42
  organomercury 35–40
  organotin 42–47
Toxic effects of metals 36
Toxicity testing 61–72

Voltammetry of organomercury 106

Waste waters
  organoarsenic in 199–201
  organolead in 151
  organomercury in 110, 111

X-ray fluorescence spectroscopy of organomercury 92

Zinc, bioaccumulation in barnacles 86, 87